Bayesian Multiple Target Tracking

Second Edition

For a complete listing of titles in the
Artech House Radar Series,
turn to the back of this book.

Bayesian Multiple Target Tracking

Second Edition

Lawrence D. Stone
Roy L. Streit
Thomas L. Corwin
Kristine L. Bell

ARTECH
HOUSE

BOSTON | LONDON
artechhouse.com

Library of Congress Cataloging-in-Publication Data
A catalog record for this book is available from the U.S. Library of Congress.

British Library Cataloguing in Publication Data
A catalogue record for this book is available from the British Library.

Cover design by Vicki Kane

ISBN 13: 978-1-60807-553-9

© 2014 ARTECH HOUSE
685 Canton Street
Norwood, MA 02062

10 9 8 7 6 5 4 3 2 1

Contents

Preface to Second Edition

The presentation and chapters in this edition have undergone substantial revision from the first edition. Much of the text and many of the discussions and examples are new. This recognizes that a lot has changed in the multiple target tracking field since 1999 when the first edition was published.

One of the most dramatic changes is in the widespread use of particle filters to implement nonlinear, non-Gaussian Bayesian trackers. Reflecting this change, we provide a detailed description of a basic particle filter that implements the Bayesian single-target recursion in Chapter 3. We describe particle filter implementations of many of the tracking algorithms derived in this book and provide numerous examples that involve the use of particle filters.

Chapter 4 on classical multiple target tracking has been substantially reorganized and now presents three algorithms that provide approximate solutions to the multiple target tracking problem, namely multiple hypothesis tracking (MHT), joint probabilistic data association (JPDA), and probabilistic multiple hypothesis tracking (PMHT). In addition, a particle filter implementation of JPDA is described.

In Chapters 5 and 6, we have included the new topics of multitarget intensity filters (iFilters) and the maximum a posteriori penalty function (MAP-PF) technique for multiple target tracking. Intensity filters are a generalization of probabilistic hypothesis density (PHD) filters.

Two chapters of the first edition (namely, Chapters 6 and 7) on likelihood ratio detection and tracking (LRDT) have been condensed into a new Chapter 7, which presents a gridded implementation of LRDT and describes the use of LRDT as a multiple target detector tracker. Examples have been added to illustrate the use of LRDT. The chapter closes with a description of iLRT, which is a combination of the iFilter and LRDT that solves the problems of determining the number of targets present and estimating their tracks. These problems have long vexed iFilters and PHD filters.

We hope the reader will find this edition broader in scope and easier to read than the first with improved motivation, new examples, and additional methods that can be applied to multiple target detection and tracking problems.

Introduction

In this book we develop the mathematical theory of multiple target tracking from a standard Bayesian point of view. The multiple target tracking problem is to estimate the number of targets in a region of interest and the state of each target.

This book views multiple target tracking as a Bayesian inference problem. Within this framework it develops the theory of single target tracking, multiple target tracking, and likelihood ratio detection and tracking (LRDT). The development emphasizes the use of likelihood functions to represent information. Likelihood functions replace and generalize the notion of contact. They provide a common currency for valuing and combining information from disparate sensors, particularly when these sensors have different measurement spaces. There is much concern, particularly in defense applications, about the problems of multiple target, multiple sensor, and multiple platform tracking. The development presented in this book provides a unified and mathematically sound approach for dealing with these problems.

While the presentation is careful and rigorous, we have strived to make it as simple and approachable as possible. We feel the standard Bayesian approach is rich and powerful enough to handle most multiple target tracking problems With this in mind we have avoided concepts such as random sets, finite set statistics, and imprecise probabilities because we feel they add unnecessary complexity and difficulty to the process of understanding multiple target tracking.

Equipped with a comprehension of the Bayesian approach and the use of likelihood functions to represent measurements and information, the reader will be able to formulate and attack single or multiple target tracking problems. The book steps the reader through this process for basic single target, multiple target, and track-before-detect problems and provides examples that illustrate practical approaches to solving these problems. Having understood this process, the reader will be prepared to tackle the myriad variations on these problems not addressed by this book.

BAYESIAN INFERENCE

What does it mean to view tracking from the viewpoint of Bayesian inference? It means that the following structure is imposed on the problem. First, we specify a prior distribution on the number of targets, their states, and how they can change state (e.g., how they move through state space). Second, when information is received in the form of sensor responses, it is converted into likelihood functions defined on the target state space. Third, likelihood functions are combined with the prior distribution using Bayes' rule to compute the posterior distribution on target state. The posterior distribution is the primary output of Bayesian inference. Sometimes it is convenient and reasonable to summarize this distribution by a single number or point estimate, but often it is not.

All questions are answered within this Bayesian framework. For example, probabilities and expected values are calculated using the standard probability calculus, and likelihood functions are presumed to contain all the relevant information in the observed data (see Chapter 2). We advocate and use the Bayesian approach because it has stood the test of time by producing consistent and reasonable answers.

The following paragraphs provide a short summary of the contents of the chapters in this book.

CHAPTER 1: TRACKING PROBLEMS

Chapter 1 describes the basic tracking problem and provides examples of four tracking problems that are representative of the problems addressed in this book. The first involves tracking a single target using a Kalman filter. The second is a bearings-only tracking problem that is solved by the use of a particle filter. The third example demonstrates the power of LRDT when applied to a problem where the Signal-to-Noise Ratio (SNR) is low and the clutter or false alarm rate is high. The forth is a multiple target tracking example.

CHAPTER 2: BAYESIAN INFERENCE AND LIKELIHOOD FUNCTIONS

This chapter presents our rationale for the use of Bayesian inference to tackle multiple target detection and tracking problems. It emphasizes the use of likelihood functions and provides a number of examples to show how they can be used to represent and combine measurements of disparate types such as position estimates, detection/nondetection events, and unthresholded sensor data.

CHAPTER 3: SINGLE TARGET TRACKING

This chapter derives the general Bayesian recursion for single target tracking when the target motion model is Markovian. This recursion produces the posterior distribution on target state at time t given the measurements received through time t. The recursion operates by using the posterior at time t and the measurements received at time $t' > t$ to compute the posterior at time t'. This chapter derives the Kalman filter as a special case of this recursion and presents a particle filter implementation of the general Bayesian recursion

CHAPTER 4: CLASSICAL MULTIPLE TARGET TRACKING

Chapter 4 begins by deriving a general Bayesian recursion for multiple target tracking. It then restricts its attention to situations in which contacts are received in scans. A contact is a detection coupled with a measurement such as a target position estimate. In each scan, there may be contacts from a number of targets as well as false detections. However, there is at most one contact from each target, and a contact is due to only one target or false detection. When contacts are received in scans, the main obstacle to successful tracking is associating contacts to targets. The chapter presents three approaches to solving this very difficult problem.

The first approach decomposes the association problem into data (contact) association hypotheses. A data association hypothesis assigns each contact in the set of contacts received by the tracker to a specific target or to a false detection. One then computes the posterior distribution on the states of the targets using this association hypothesis. By computing the probability that each data association hypothesis is correct, one can identify the highest probability one and display the target state distributions resulting from it. This is called multiple hypothesis tracking (MHT). Under certain independence assumptions, this chapter derives an independent MHT recursion. In the case where the target motion model is Gaussian and contacts are linear functions of target state with Gaussian errors, a special linear-Gaussian MHT recursion is obtained.

Approximations are required to implement MHT recursions. In particular, one must limit the number of hypotheses to prevent the computations from increasing in an exponential fashion as the number of contacts increases.

In MHT a data association hypothesis requires that each contact be assigned to a target or false detection with probability $p = 1$ or 0. Joint probabilistic data association (JPDA) relaxes this "hard" assignment assumption by allowing association probabilities $0 < p < 1$. It assumes the probability distributions computed for target state at time t are a sufficient statistic for computing target state distributions from contacts obtained after time t. Under these assumptions, the chapter derives the JPDA recursion for multiple target tracking. This recursion avoids the exponential increase in computational load that would result from an MHT algorithm that does not limit the number of hypothesis. A particle filter implementation of JPDA is also provided.

The third method, probabilistic multiple hypothesis tracking (PMHT), employs still another approximation for computing multiple target tracking solutions. It relaxes the assumption that at each time period each target generates at most one contact. This leads to a multiple target tracking algorithm whose computations increase linearly with the number of contacts and targets.

CHAPTER 5: MULTITARGET INTENSITY FILTERS

Intensity filters take a different approach to multiple target tracking than presented in Chapter 4. Intensity filters do not explicitly consider the association of contacts to targets. Instead they compute intensity functions that estimate the spatial density of targets. Peaks in this function indicate target locations. This greatly simplifies the computations involved in multiple target tracking but at the expense of not explicitly producing target state estimates or target tracks. This chapter derives an intensity filter called the iFilter. For this filter, the standard target state space is augmented to include a clutter target state that generates clutter or false detections. Multiple target distributions are modeled by the use of Poisson point processes (PPPs). The iFilter provides a recursive method of computing the intensity function of the posterior point process at time t given the contacts received through time t. The recursion proceeds by approximating the (Bayes) posterior point process by a PPP whose intensity function is calculated by the iFilter. The addition of a clutter target state allows the iFilter to estimate the false detection rate dynamically as a function of the contacts received. The probability hypothesis density (PHD) filter, which is restricted to the standard target state space and uses a fixed prior estimate of the rate of false detections, is derived as a special case of the iFilter.

CHAPTER 6: MULTIPLE TARGET TRACKING USING TRACKER GENERATED MEASUREMENTS

Chapter 6 drops the assumption that observations are provided to the tracker in the form of contacts and assumes that one has access to unthresholded sensor data. It develops a multiple target tracking method called the maximum a posteriori probability penalty function (MAP-PF) technique. MAP-PF jointly performs target state estimation and the signal processing to obtain measurements. Using estimates of target state for each target, MAP-PF generates one measurement for each target for each scan. This renders moot the association problem that vexes most multiple target trackers. The measurements are used to update target state distributions and provide state estimates at the time of the next scan. MAP-PF requires an outside mechanism to estimate the number of targets present. One possibility is to use LRDT as described in Chapter 7 to perform this function.

CHAPTER 7: LIKELIHOOD RATIO DETECTION AND TRACKING

This chapter explores the problem of detection and tracking when there is at most one target present. It develops LRDT based on an extension of the single target tracking methodology presented in Chapter 3. However, the methodology also works well for certain multiple target detection and tracking problems. The chapter presents the theory and basic recursion for LRDT and shows by examples how it can be successfully applied to low SNR and high clutter rate situations to produce high detection probabilities with low false alarm rates. A grid-based implementation of LRDT is presented.

The chapter closes with a description of iLRT, which is a combination of multitarget intensity filtering presented in Chapter 5 and LRDT presented in this chapter. This combination provides a solution to two problems that confront intensity or PHD filtering, namely estimating the number of targets present and producing track estimates for the identified targets.

Acknowledgments

The authors would like to acknowledge the help they received in producing the examples in this book. In particular we thank Stephen L. Anderson for producing Examples 2 and 4 in Chapter 1, Michael A. Hogye for producing the multistatic sonar example in Section 7.5.2, and Bryan R. Osborn for the iLRT example in Section 7.8.3.

Chapter 1

Tracking Problems

The purpose of this chapter is to acquaint the reader with tracking problems and provide some representative examples. Section 1.1 describes the basic tracking problem. Sections 1.2–1.4 present examples of single target tracking problems, and Section 1.5 presents a multiple target tracking example. The examples reflect the authors' background in naval applications, but the principles illustrated apply to all tracking problems. Some readers may want to absorb the basic ideas presented in the examples on first reading and return to examine them in detail after reading Chapters 3, 4, or 7. The examples illustrate the concepts presented in those chapters.

1.1 DESCRIPTION OF TRACKING PROBLEM

A typical tracking problem involves estimating the path of an object (target) of interest such as a plane, car, or ship. There is a sensor such as a radar that provides measurements of the position of the target at a discrete set of times, $t = t_0 < t_1 < \ldots < t_K$. For convenience we will set $t_0 = 0$. If the radar measurements, typically range and bearing from the sensor location, have no error in them, then we can simply "connect the dots" to estimate the target's path. However, in the problems of interest to us, the measurements have unknown errors. In fact some of the measurements may be spurious in the sense that they are not generated by the target of interest or that they are false alarms generated by noise or clutter.

For the moment, let us presume that all measurements are generated by the target of interest. Using these measurements, we want to estimate the target's position at the present time $t = t_K$, its path over $[0, t_K]$, or its position at some time $t > t_K$. In the estimation literature (tracking is special case of estimation), estimating the present position is *filtering*, estimating the future position is *prediction*, and estimating the past path or positions is *smoothing*. More generally, we may want to estimate other characteristics of the target such as its velocity or acceleration, so we define the target *state* as the vector of target components that we want to estimate (e.g., position and velocity). The process of estimating a target's state from noisy measurements is called *tracking*.

1.1.1 Measurement and Motion Models

To perform tracking, we need mathematical models. First we need a model for the measurement error. This model is given in terms of a probability distribution for the error in a measurement. Second, we need a model for the target's motion. A simple example is a constant velocity model. For this model we could specify a probability distribution on the target's position at time $t = 0$ and a distribution on the target's velocity. If the target's initial position and velocity were known, then its complete path would be known. More generally, the target motion model is a stochastic process that describes in a probabilistic sense how the target moves through its state space.

1.1.2 Estimation

For single target tracking we wish to combine the measurement data with the measurement and motion models to estimate the target's track. More specifically, we wish to obtain an optimal or best estimate, but what do we mean by optimal? One criterion could be to minimize the expected squared error between the estimated and true target track. Another might be to find the track with maximum probability of being the correct one. Still another could be to compute the posterior distribution on target state given the measurements. In this case, the stochastic process modeling target motion forms the prior, and the measurements are the data used to compute the posterior (i.e., the conditional distribution on the target motion model given the measurements at $t = 0 < t_1 < \ldots < t_K$). This is the Bayesian point of view, which is taken in this book. In this view, the posterior contains all the information available about the target state. We use the term Bayesian inference to denote the computation of the posterior distribution and any estimates derived from it such as the mean or the maximum a posteriori probability (MAP) state. The mean of the posterior is the estimate that minimizes the expected squared error between the estimated and actual target state.

1.1.3 Filters

The best known and most prevalent method for solving tracking problems is the Kalman filter [1]. This was originally developed as an iterative least squares method of solving filtering (tracking) problems without the Bayesian interpretation given in this book. We employ the Bayesian interpretation, because it allows us to place the Kalman filter in a larger class of Bayesian tracking methods and provides additional insight into the method itself.

1.1.3.1 Kalman Filtering

The Kalman filter applies to situations where the target motion process is Markov and Gaussian. In a Gaussian process, the distribution on target state at any time t is Gaussian, and the multivariate distribution on the target states at any finite set of times is multivariate Gaussian. In addition the Kalman filter requires that the measurements be linear functions of the target state at the time of the measurement with additive Gaussian errors. A further assumption is needed. The measurement errors must be independent of target state at the time of the measurement and of the other measurement errors. This set of assumptions is called the *linear-Gaussian* assumptions.

In the linear-Gaussian case, the posterior distribution on target state at any time t is Gaussian. A Gaussian distribution is characterized by its mean and covariance. The Kalman filter provides a simple and efficient recursion for calculating the mean and covariance of the Gaussian posterior target state distribution at the present time t_K given the measurement at t_K and the mean and covariance of the posterior target state distribution at t_{K-1}. In this case, the mean of the Kalman filter solution minimizes the expected squared error between the estimated target state and the true one and is the MAP estimate of target state. In the linear-Gaussian case, the Kalman filter solves for all three best estimates defined above. It finds the posterior distribution on target state and computes the mean of that distribution, which minimizes the expected squared error and maximizes the a posteriori probability.

If we relax the Gaussian assumptions but retain the linear and independence assumptions, the Kalman filter provides the optimum estimate of target state that *is a linear function of the measurements*. This estimate is optimum in the sense of minimizing the expected squared error.

1.1.3.2 Extensions of Kalman Filtering

There are many extensions and variations on the Kalman filter that are employed when the linear-Gaussian assumptions do not hold. In cases where measurements are nonlinear functions of target state or the equations defining target motion are nonlinear, extended Kalman filters (EKFs) replace these equations with linear approximations and then use a Kalman filter to obtain an approximate solution [2–4]. Unscented Kalman filters approximate non-Gaussian distributions on target state and measurement errors by calculating an approximate mean and covariance of these distributions and replacing the actual distributions with Gaussian ones having the calculated means and covariances [2, 5, 6]. Another Kalman-related approximation is the Gaussian sum filter. In this case non-Gaussian distributions are approximated by weighted sums of Gaussian distributions allowing a Kalman filter to be applied to each of the Gaussian components [7, 8]. There are extensions such as the interacting multiple model (IMM) Kalman filter where the Gaussian motion model is replaced by a distribution on Gaussian motion models [9]. We do not explore these extensions. This would take us far afield from our purpose of presenting tracking from a Bayesian point of view. The reader can consult Chapter 2 of [2] for a summary of these methods. In addition, these methods are discussed in [3, 10, 11].

1.1.3.3 Nonlinear Filtering

When the linear-Gaussian assumptions do not hold, we are in the situation where the motion model and measurement error distributions may be non-Gaussian, and the relationships between target state and measurements may be nonlinear. This is referred to as *nonlinear* tracking. It would be more accurate to say nonlinear, non-Gaussian tracking, but the term nonlinear is generally used to mean both nonlinear and non-Gaussian. With the exception of some special cases (see [12–14]), there is no analytic recursion that can be used to solve nonlinear tracking problems. Instead we are forced to use discrete approximations to perform the filtering calculations. There are two types of approximations. The first approach is a grid-based approximation where the target state space is approximated by a discrete grid of cells (see [15, 16]). In this case the motion model is approximated by a discrete state Markov process where the states correspond to the grid cells. The transition probabilities for the chain are designed to approximate the target motion through the state space. In addition, the relationships between target state and measurements must be converted to this discrete state space. In order to preserve the resolution of a grid-based tracker, the set of grid cells must be dynamically recentered and rescaled as the target state distribution translates, spreads out, or condenses. Chapter 3 of the first edition [17] of this book describes a grid-based tracker in detail.

The second approach is to approximate the stochastic process that models target motion by a discrete set of sample paths drawn from the process. The sample paths are called particles and the method particle filtering. As with the grid-based approach, one needs a process to retain resolution when the probability distribution on target state condenses. For particle filters, this process is resampling [2] and [18], which is discussed in the bearings-only example in Section 1.3 and in Chapter 3.

The grid-based approach suffers from two difficulties. First dynamic regridding is complex to implement in computer code, and second, the use of grids tends to be inefficient in the sense that many grid cells will contain little or no probability. In addition, the memory and computational requirements become overwhelming as the dimension of the state space increases beyond four. By contrast, particle filters are relatively easy to implement on a computer, and the resampling process guarantees that computational effort will be expended more efficiently. In particular it will be expended only on particles with positive probability. In addition, particles with very low probability tend to be "killed-off," further increasing the computational efficiency. For these reasons, the discussion of numerical implementation of nonlinear filters in this edition is limited to particle filters.

1.2 EXAMPLE 1: TRACKING A SURFACE SHIP

This example involves tracking a surface ship. Even though the automated identification system (AIS) is used on many ships, tracking surface ships in the open ocean remains a challenging problem. AIS is an automatic tracking system in which location obtained from global positioning system (GPS) and ship identification information is broadcast to nearby ships and AIS base stations. When AIS information is available, tracking is not a problem. However, maintaining surveillance on ships in the open ocean remains an important naval problem, because often there are no AIS base stations in position to receive the AIS broadcast information and because military ships do not broadcast their positions on AIS. In addition, some commercial ships do not broadcast AIS information, because they do not wish to be tracked. In these cases one must rely on long-range radar and overhead sensors to track surface ships and maintain surveillance on targets of interest.

We consider a simplified situation in which there is a single target in the area of interest, and there are no false alarms. There is a satellite sensor that provides position measurements with Gaussian errors at irregular time increments. The state space for the target is four-dimensional with two position dimensions in the normal (x_1, x_2) Cartesian coordinate system and two velocity coordinates, (v_1, v_2). We follow the convention that velocity is a vector quantity and that speed is the magnitude of the velocity vector.

For this example we need some facts about multivariate Gaussian distributions. An r-dimensional Gaussian distribution is defined by its r-dimensional mean μ and covariance matrix Σ. We use $\mathcal{N}(\mu, \Sigma)$ to denote this distribution. It has probability density function f on the r-dimensional space $\mathbb{R}^r$ where

$$f(x) = (2\pi)^{-\frac{r}{2}} |\Sigma|^{-\frac{1}{2}} \exp\left(-\frac{1}{2}(x-\mu)^T \Sigma^{-1}(x-\mu)\right) \text{ for } x \in \mathbb{R}^r \qquad (1.1)$$

The vector x is a column vector and x^T denotes the transpose of x.

The equiprobability density contours for this distribution are ellipsoids. The ellipsoids are defined by the set of points x such that

$$(x-\mu)^T \Sigma^{-1} (x-\mu) = c^2 \text{ for some } c > 0 \qquad (1.2)$$

In two-dimensional space, this set of points is an ellipse and (1.2) defines the c-σ uncertainty ellipse.

1.2.1 Prior Distribution on Target State

For this example the prior distribution on target state is four-dimensional Gaussian. Figure 1.1 shows the distribution on position, which is circular Gaussian with mean (2, 28) and variance $\sigma^2 = 40$ nautical miles squared in any direction. The axes in the figure give distances in nautical miles (nm) from the origin. The solid circle shows the 2-σ uncertainty ellipse for this distribution. The marginal velocity distribution is bivariate Gaussian with mean $(2, -6)$. The mean velocity is represented by a vector with its tail placed at the mean of the position distribution. The length of the vector is equal to 6.3 nautical miles per hour, knots (kn), which is the speed of the mean velocity. (When we are plotting velocity vectors in Figure 1.1 and subsequent figures, the axes are to be interpreted as measuring knots in the x_1 and x_2 directions.) The 2-σ uncertainty ellipse for the velocity distribution is shown by a solid line centered on the tip of the mean velocity vector. The velocity distribution shows more uncertainty in the x_1 direction than in the x_2 direction.

This prior distribution represents a situation where we have some knowledge of both the ship's position and its velocity. If our knowledge is very uncertain, we can represent that by a uniform distribution over a large region in both position and velocity space, which can be approximated by Gaussians with large variances.

1.2.2 Motion Model

In addition to specifying the initial distribution, we must specify a motion model for the target. In this case we use the integrated Ornstein-Uhlenbeck (IOU) process described in Section 3.2.2.2. The target's velocity is modeled as a stationary Gaussian process with an exponential autocorrelation function, $e^{-\gamma|t|}$, for all times t. It is initialized by a bivariate Gaussian distribution. Since the velocity process is Gaussian, the target's velocity distribution is Gaussian at any time. The target changes its velocity in a probabilistic fashion, and the parameter γ in the autocorrelation function determines the rate at which the velocity changes occur. In somewhat loose terms, we can think of $1/\gamma$ as the mean time between velocity changes. The parameter $1/\gamma$ is often called the decorrelation time. These terms are suggestive but somewhat inaccurate. Along a sample path, the target's velocity is changing continuously, and its correlation with a previous velocity decreases exponentially as time increases. The sample paths of the velocity process are integrated to produce the sample paths for the movement of the target from its initial position.

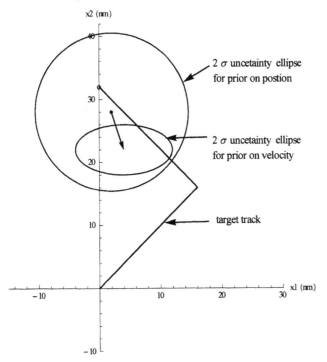

Figure 1.1 Prior target state distribution at time 0. The vector shows the mean of the prior velocity distribution. The tail is on the mean of the prior position distribution. The letter "o" marks the actual position of the target at time 0.

For this example we have chosen $\gamma = 0.075$ /hr so that the "mean time" between velocity changes is 13.3 hr. The IOU process has one other parameter, σ_0. As time increases, the limiting velocity distribution of the process is circular Gaussian with mean 0 and variance $\sigma_0^2 / 2\gamma$ along each axis. This distribution has a mean squared speed of σ_0^2 / γ. For this example, we have chosen σ_0 to yield a root mean square speed of 5.8 kn for the limiting velocity distribution. This produces a convenient Gaussian motion model for a merchant ship. However, is it more realistic to model merchant ships as moving along rhumb[1] lines or great circle arcs. This is easily done with a particle filter but is difficult with a Kalman filter.

---

[1] A rhumb line is the path that results from following a constant heading. Headings are measured in degrees clockwise from north. For example, a heading of 270° is due west. A rhumb line does not follow a great circle route except for ones heading due north/south or heading due east/west along the equator. A rhumb line is easier to navigate than a great circle route, which generally requires continual heading changes.

1.2.3 Measurement Model

In this example, the sensors (satellites) obtain estimates (measurements) of the ship's position at irregular time intervals. The measurement errors have bivariate Gaussian distributions with mean $(0,0)$ and covariance matrices, which may vary from measurement to measurement as the satellite viewing angle changes. In mathematical terms, the measurement Y_k at time t_k is related to the target state $X(t_k)$ at time t_k as follows

$$Y_k = HX(t_k) + \varepsilon_k \tag{1.3}$$

where

$$X(t_k) = \left(x_1, x_2, v_1, v_2\right)^T, \; H = \begin{pmatrix} 1 & 0 & 0 & 0 \\ 0 & 1 & 0 & 0 \end{pmatrix}, \; \varepsilon_k \sim \mathcal{N}(0, \Sigma_k)$$

The first two components of $X(t_k)$ are the position components. H is called the measurement matrix. The notation $\varepsilon \sim \mathcal{N}(\mu, \Sigma)$ means that ε has a Gaussian distribution with mean μ and covariance Σ. Another way to express (1.3) is $Y_k \sim \mathcal{N}(HX(t_k), \Sigma_k)$.

The measurement at time t_k equals the position of the target at time t_k plus a bivariate Gaussian error with mean $(0,0)$ and covariance Σ_k. Note that the relationship between the measurement and the target state is linear and that the measurement error is Gaussian. These assumptions are crucial for Kalman filtering. We also assume that ε_k is independent of $X(t_k)$ and ε_j for $j \neq k$.

1.2.4 Tracker Output

We simulated measurements using the above model and applied the continuous-discrete Kalman filter recursion given in Section 3.2.2 to obtain the tracker output for this example. Observations were obtained at 0, 2, 3, 4.5, 5.5, 6, 7, and 8 hours. (Times are measured from the time of the initial distribution.) Figure 1.2 shows the results of processing the first four observations. The figure for time 0.0 hr shows the 2-σ uncertainty ellipse (circle) for the prior position distribution as the larger solid ellipse (circle). In a bivariate Gaussian distribution, the 2-σ ellipse contains 86% of the probability mass. The dashed ellipse is the 2-σ uncertainty ellipse for the measurement uncertainty distribution. The smaller solid ellipse is the 2-σ uncertainty ellipse for the posterior position distribution after incorporating the measurement at time 0. The vector with its tail placed on the dot at the center of this ellipse is equal to the mean of the posterior velocity distribution. The two solid straight lines show the path of the target over the eight hours of the example. The letter "o" marks the position of the target at the time of the figure. The target's actual speed is 5.7 kn. The speed corresponding to the posterior mean velocity vector is listed as the mean speed, which is 6.3 kn for time 0. Since we have only one position observation, and it is at time 0, the posterior mean velocity is the same as the mean velocity of the prior distribution.

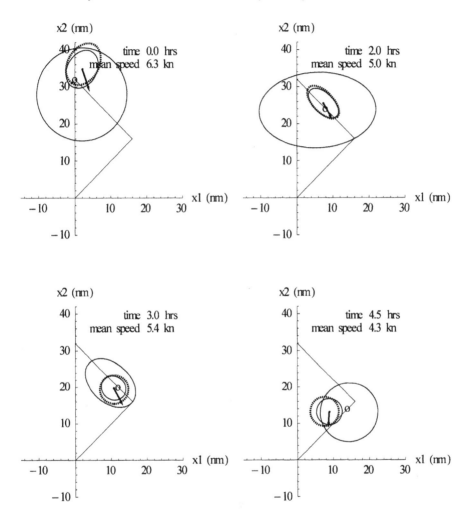

Figure 1.2 Tracker output at the first four measurement times: The polygonal line shows the target path with its position on the path marked by the letter "o." For each time, the large solid ellipse is the 2-σ uncertainty ellipse for the motion updated target position distribution. The dashed ellipse is the 2-σ uncertainty ellipse for the measurement. The smaller solid ellipse is the 2-σ uncertainty ellipse for the posterior target position distribution given the measurement. The vector is the mean velocity, and the mean speed is the magnitude of this vector. The tail of the vector is on the mean of the posterior distribution.

Since we have specified the prior distribution on target state, we are able to compute a posterior on the full target state even though our first measurement does not include all of the target state components. Since the prior is Gaussian and the measurement is a linear function of target state with a Gaussian error, the posterior is also Gaussian.

In Figure 1.2, the larger solid ellipse for time 2.0 hr shows the position distribution from time 0 updated for target motion to time 2.0 hr. This is the motion updated or predicted target state distribution at time 2.0 hr before the measurement. Note that the measurement at time 2.0 hr has different error characteristics (uncertainty ellipse) than the measurement at time 0.0 hr. In addition, the posterior mean velocity vector is now different from that of the initial distribution and better aligned with the actual target velocity.

At time 4.0 hr, the target changes velocity. Note that the actual target path does not follow the IOU motion model. As mentioned above, the IOU model is convenient for a Kalman filter but is not a good representation of the actual motion of ships. A measurement is received at time 4.5 hr, which has considerable error in it. The actual target position is slightly beyond the edge of the 2-σ ellipse of the posterior distribution, but the mean velocity is beginning to move around to the target's new velocity. Since the 2-σ ellipse contains only 86% of the probability distribution, we should expect the actual target position to lie outside this ellipse 14% of the time.

Figure 1.3 shows the tracker output resulting from the last four observations. At 6.0 hr the tracker has a poor estimate of the target's velocity. This results from the target's recent course change and a measurement at 6.0 hr with a large error. This produces a predicted target distribution at 7.0 hr that has the target's actual position on the edge of the distribution. The measurement at 7.0 hr, which has a small error, pulls the mean of both the position and velocity distribution close to that of the target. At 8.0 hr, the predicted target distribution is greatly improved, and the posterior position and velocity distributions are consistent with the target's position and velocity.

A more detailed discussion of Kalman filtering may be found in [2, 3, 10, 11]. The example given above represents a high data rate (compared to the target's speed) and high SNR situation. The measurements come often, and there are no false alarms. The relationships between target state and the measurements are linear with Gaussian measurement errors, and the motion model produces Gaussian distributions. This is the case in which the Kalman filter is optimal. However, this optimality must be qualified by remembering that the motion model is not a realistic one for a ship.

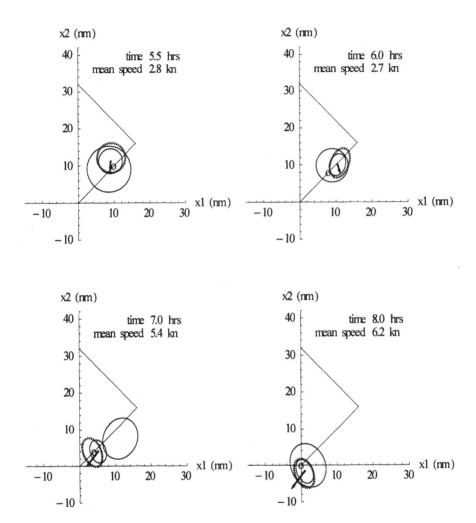

Figure 1.3 Tracker output at the last four measurement times. The notes in the caption of Figure 1.2 identify the ellipses, polygonal path, and the vector that appear at each time.

1.3 EXAMPLE 2: BEARINGS-ONLY TRACKING

If one submarine is tracking another using a passive acoustic sensor, the measurements are typically noisy estimates of the bearing of the target from the sensor with limited range information. The process of going from a series of bearing measurements made over time to an estimate of the target's range, course, and speed is called target motion analysis (TMA). Since these tracking problems tend to take place over relatively short distances, it is reasonable to view the target motion as taking place in the plane. The TMA problem is a particularly vexing one for a Kalman filter, because the measurements are not linear in the natural target state, namely position and velocity in the plane. In addition, although the measurement errors may be Gaussian in bearing, the transformation of these errors to the plane produces distinctly non-Gaussian error distributions.

For those familiar with bearings-only tracking and observability problems, the particle filter method described in this example deals with these issues in a natural and effective manner.

Since the measurements are not linear in target state and the errors are not Gaussian, a Kalman filter does not apply directly. There have been many attempts to approximate the problem so that it can be solved by a Kalman filter. A typical approach is to choose an estimated range for the target and to use that range to produce a linear approximation to the bearing measurement along with a Gaussian approximation to the measurement errors. This is called an EKF. Kalman filter approximations have difficulties such as range collapse, the tendency of the filter solution to "collapse" to zero range from the sensor.

In this example we demonstrate how a simple, nonlinear particle filter can be used to solve the problem of bearings-only tracking of a maneuvering target. In [2] the performance of particle filters for bearings-only tracking has been compared to numerous Kalman filter approximations against maneuvering and nonmaneuvering targets. The performance of the particle filters was shown to be superior to the Kalman filter approximations. A perusal of Chapter 6 of [2] reveals how complex and mathematically daunting these approximations can become. All to no avail, the simple particle filter outperforms them all.

The example given here and the approach to particle filters employed in this book differ in some significant respects from [2]. First, we tend to use the sampling importance resampling (SIR) particle filter described in Section 1.3.5 in this example. We use this filter, because it follows the natural Bayesian recursion given in Chapter 3 and performs well in most situations. Second, we view the particles as sample paths drawn from the stochastic process that forms the target motion model. The probabilistic nature of the process represents our uncertainty about the target's motion or behavior. This produces more natural and realistic motion models than an approach that views target motion as deterministic with process noise added to account for mismodeling or unforeseen circumstances. The example that follows illustrates our approach as well as its power and simplicity.

1.3.1 Description of Example

Figure 1.4 shows the paths of the tracking submarine (ownship) and the target submarine. Ownship maneuvers early in its track before the target submarine does. Later the target submarine maneuvers and after that ownship maneuvers again. Measurements were generated at a sequence of times $t_1, t_2, \ldots$ spaced one minute apart. (Note that measurements do not need to be equally spaced.) The measurements were simulated by adding a Gaussian error to the actual bearing of the target from ownship.

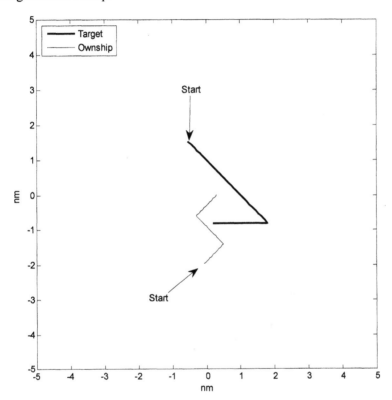

Figure 1.4 Paths of target and ownship submarines.

1.3.2 Prior Distribution

The prior on position at time 0 is uniform over the rectangle shown in Figure 1.4. The target's prior velocity is given by assuming a uniform distribution for target heading and an independent uniform distribution on speed between 2 and 20 kn. The prior velocity distribution is contained in the annulus shown in Figure 1.5 where $S_{min} = 2$ kn, $S_{max} = 20$ kn, and $v = (v_1, v_2)$ is velocity. The minimum speed derives from the fact that submarines require a minimum speed to maintain their ability to steer and balance the submarine. The maximum speed constraint is natural for a submarine that expects it may be in the vicinity of another submarine. Note that this velocity distribution is not well approximated by a Gaussian one.

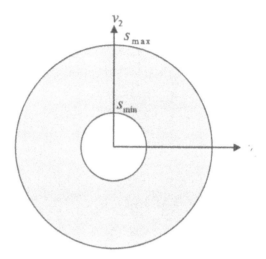

Figure 1.5   Prior velocity distribution annulus

1.3.3 Motion Model

Target motion is modeled by choosing an initial position and velocity from the priors described above. The target makes velocity changes at times that are exponentially distributed. A random time τ has an exponential distribution if

$$\Pr\{\tau \leq t\} = 1 - e^{-\lambda t} \text{ for } t > 0 \text{ where } \lambda > 0$$

The mean of this distribution is $1/\lambda$. If the target is in state $(x(t_0), v(t_0))$ at time t_0, it continues at velocity $v(t_0)$ for a random time τ that is exponentially distributed with mean ½ hour ($\lambda = 2/\text{hr}$). Thus

$$(x(t), v(t)) = (x(t_0) + (t - t_0)v(t_0), v(t_0)) \text{ for } t_0 \leq t \leq t_0 + \tau$$

At time $t_0 + \tau$, a velocity change Δv is obtained by independent draws for the change $\Delta\theta$ in heading and change Δs in speed. The change in heading $\Delta\theta$ is drawn from distribution D_1 with probability 0.5 and distribution D_2 with probability 0.5 where

$$D_1 = \mathcal{N}(\mu_h, \sigma_h^2) \qquad \text{truncated at zero (no negative values)}$$
$$D_2 = \mathcal{N}(-\mu_h, \sigma_h^2) \qquad \text{truncated at zero (no positive values)}$$

The heading changes are further constrained to be between -180 and 180 degrees. Figure 1.6 shows an example of the density for a course change distribution.

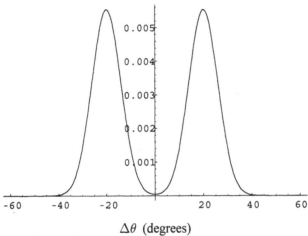

$\Delta\theta$ (degrees)

Figure 1.6   Course change density

The change in speed Δs is drawn from distribution D_1' with probability 0.5 and distribution D_2' with probability 0.5 where

$$D_1' = \mathcal{N}(\mu_s, \sigma_s^2) \qquad \text{truncated at zero (no negative values)}$$
$$D_2' = \mathcal{N}(-\mu_s, \sigma_s^2) \qquad \text{truncated at zero (no positive values)}$$

The speed changes are further truncated so that the resulting speed does not fall below $S_{\min}$ or above $S_{\max}$. This produces a speed change density with a shape similar to the one in the figure above.

For this example, we chose,

$$\mu_h = 60\,\text{deg} \quad \sigma_h = 30\,\text{deg}$$
$$\mu_s = 2\,\text{kn} \qquad \sigma_s = 1\,\text{kn}$$

Clearly, submarines do not change velocity instantaneously, so the model is an approximation in that sense. Since we will be simulating the target paths, we could round the edges if we chose and put a constraint on the turn radius. We could also add circular arcs if we wished. Our experience is that this is not necessary for good performance in this problem.

The process of choosing the motion model for the target submarine in this example is instructive. Because we are using a particle filter, we can concentrate on modeling the operational and tactical constraints of the submarine. How slow or fast will it go? How often will it change velocity? Is there a general direction of motion? We represent our uncertainty about the behavior of the submarine by probability distributions. This provides for natural and realistic (to the appropriate level of detail) models for target motion. The resulting simulation of the target's motion becomes the motion model for the particle filter.

1.3.4 Measurement Model

Suppose that $(y_1(t_k), y_2(t_k))$ and $(x_1(t_k), x_2(t_k))$ are the positions of ownship and the target submarine at time t_k of the kth bearing measurement θ_k. The measurement θ_k satisfies the equation[2]

$$\theta_k = \arctan\left(\frac{x_2(t_k) - y_2(t_k)}{x_1(t_k) - y_1(t_k)}\right) + \varepsilon_k \text{ where } \varepsilon_k \sim \mathcal{N}\left(0, \sigma_k^2\right) \tag{1.4}$$

That is, the observed bearing of the target from ownship has a Gaussian error in bearing with mean 0 and standard deviation σ_k, which may depend on the measurement. For this example we have taken $\sigma_k = 2 \deg$.

This nonlinear measurement with non-Gaussian error in the standard (x_1, x_2) coordinate system is easily incorporated into a particle filter by the use of a likelihood function. The likelihood function L_1 for the measurement is the probability (density) of obtaining measurement θ_k as a function of target position (x_1, x_2). From (1.4) we calculate L_1 as follows

$$L_1\left(\theta_k \mid (x_1, x_2)\right) = \Pr\left\{\varepsilon_k = \theta_k - \arctan\left(\frac{x_2 - y_2(t_k)}{x_1 - y_1(t_k)}\right)\right\}$$
$$= \eta\left(\theta_k - \arctan\left(\frac{x_2 - y_2(t_k)}{x_1 - y_1(t_k)}\right), 0, \sigma_k^2\right) \tag{1.5}$$

where $\eta(\cdot, \mu, \Sigma)$ is the probability density function for $\mathcal{N}(\mu, \Sigma)$. Note that the measurement θ_k is data; it is fixed. As a result, L_1 is a function of the position component (x_1, x_2) of the target state space. L_1 specifies the likelihood of the measurement as a function of target position as shown in Figure 1.7.

---

[2] Although the notation does not indicate this, we are using the two-argument arctan function that accounts for the quadrant of the bearing. For mathematical convenience, the bearings in this example are measured counter-clockwise from the horizontal axis (axis 1). This differs from the usual maritime convention where bearings are measured clockwise from north (the vertical axis).

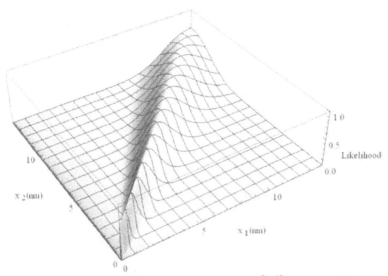

Figure 1.7 Bearing likelihood function: The sensor is located at $(0,0)$ and the bearing measurement is 45° with a 5° standard deviation for the Gaussian measurement error.

For this example we assume that the detection capabilities of the submarine sensor limit detections to no more than 6.5 nm from ownship. We model this by setting the bearing likelihood to be 0 beyond 6.5 nm from the position of ownship. If one has more detailed detection information, it can be incorporated into the likelihood function by the method described in Section 2.3.5.

On some submarines there is a towed array of passive acoustic sensors that trails a substantial distance behind the submarine. In this case one may obtain bearing measurements from the towed array as well the sensor onboard the submarine. Using likelihood functions and a particle filter, it is easy to incorporate measurements from this second sensor into the tracker even though it is not colocated with first sensor, and its measurement times may not coincide with those of the first sensor. The likelihood function for these measurements is the same as in (1.5) except that $(y_1(t_k), y_2(t_k))$ is replaced by the position of the array $(z_1(t), z_2(t))$ at the time t of the measurement θ', and σ_k is replaced by the standard deviation σ of the bearing error for the towed array.[3] Specifically

$$L_2\left(\theta' \mid (x_1, x_2)\right) = \eta\left(\theta' - \arctan\left(\frac{x_2 - z_2(t)}{x_1 - z_1(t)}\right), 0, \sigma^2\right)$$

---

[3] Towed arrays produce three-dimensional conical angle measurements as discussed in Chapter 2. These are usually approximated by two-dimensional bearings in the plane through the array parallel to the ocean surface. This produces two "mirror" bearings, one on each side of the array.

1.3.5 Particle Filter Description

The particle filter used for this example is the SIR filter described in [2]. This is a simple particle filter that gives good results for this example and serves well to explain the concept and use of a particle filter. In some situations, more complicated particle filters are required. See Chapter 3 of [2] for a more thorough discussion of particle filters.

1.3.5.1 Generating Particles

We use $N = 25,000$ particles for this example. It may seem that 25,000 is a lot of particles, but bearings-only tracking is a difficult problem. In order to obtain a good statistical representation of the joint four-dimensional position-velocity distribution, a large number of particles is required. On the other hand 25,000 particles present a modest challenge for today's computers. For the multiple target example presented in Section 1.5, we use 10,000 particles per target. The smaller number is satisfactory in this case because the measurement information, position estimates with bivariate normal errors, provides both range and bearing information.

We could generate these particles by making 25,000 independent draws for the initial position of the particles from the bounding rectangle shown in Figure 1.4 and then making independent draws for velocity of each particle from the distribution given in Section 1.3.2. However, since the posterior position distribution after the first measurement will be a normalized version of the likelihood function in (1.5) truncated at 6.5 nm from ownship, we started the particle filter at the time of the first measurement by drawing 25,000 positions from this distribution and then drawing the velocity for each particle from the initial velocity distribution. In addition we drew times from an exponential distribution with mean 0.5 hours for the time to the next velocity change for each particle. Each particle is given a weight equal to 1 indicating that it carries probability 1/25,000.

1.3.5.2 Motion and Measurement Update

When the next measurement θ_2 is received at time t_2, we motion update each particle to its position at time t_2 using the motion model in Section 1.3.3. Let $x^j(t_2)$ be the position of the jth particle at t_2. Because of the motion model, some particles will have maneuvered (possibly more than once) by time t_2 and others will have not. Next we compute

$$w^j = L_1\left(\theta_2 \mid x^j(t_2)\right) \text{ for } j = 1,...,25,000 \qquad (1.6)$$

This updates the weight of each particle to account for the measurement. The posterior probability on particle j given the observation is computed by

$$\tilde{p}^{j} = \frac{w^{j}}{\sum_{j'=1}^{25000} w^{j'}} \text{ for } j = 1,\ldots,25,000$$

This produces a discrete approximation to the posterior distribution on target state at time t_2 given the first two measurements. It is a four-dimensional distribution having two space and two velocity dimensions.

1.3.5.3 Resampling

For numerical reasons, it is a good idea to resample the particles after each measurement update. If one does not do this, probability will tend to concentrate on a small number of the particles and the effectiveness of the filter will degrade. To avoid this we use a resampling process very similar to the one called splitting and Russian roulette (see [19]), which we now describe. (The actual one we use is discussed in Chapter 3.) Resampling is a process that goes back to the earliest days of computer implementation of Monte Carlo methods.

For each particle compute $M(j) = 25000\tilde{p}^{j}$ and write $M(j) = k_j + f_j$ where k_j and f_j are the integer and fractional parts of $M(j)$. For each particle, draw a random number r_j from the uniform distribution on $[0,1]$. If $r_j < f_j$, set $k_j = k_j + 1$; otherwise leave k_j as is. Generate k_j almost identical copies of particle j. The state of each of these particles is equal to the state of particle j plus a small perturbation. Give each of the new particles weight $1/25,000$.

One can see that particles with high probability will get split into a number of almost identical particles. Correspondingly, particles with low probabilities will tend to be "killed off." This produces a representation of the posterior distribution where all particles have equal weight, and high probability density regions are represented by high concentrations of particles. When the next measurement is received we perform the above process again starting with the motion update step.

1.3.5.4 Measurements from a Second Sensor

If we have a second sensor (in the same or different location from the first) that produces a measurement θ' at the same time as first sensor, we can modify (1.6) by first computing w^j as shown there and then computing

$$w_+^{j} = w^{j} L_2\left(\theta' \mid x^{j}(t_2)\right)$$

We then compute the posterior probabilities $\tilde{p}^j$ using w_+^j in place of w^j. If the measurement from the second sensor occurs at a different time, then we simply motion update to the time of the measurement and apply the L_2 likelihood function as in (1.6). Notice that second sensor does not have to provide bearing measurements. It can provide any type of measurement so long as it can be converted to a likelihood function.

1.3.6 Comments

The recursive process described above is deceptively simple, yet (as we show in Chapter 3) it produces the Bayesian posterior distribution on target state given the measurements. It is also general and powerful. We can substitute any motion model that we can simulate for the one used above. We can incorporate measurements from any sensor for which we can calculate a likelihood function. We can incorporate information from multiple sensors, colocated or not, and the sensors can be disparate producing different types of measurements.

1.3.7 Tracker Output

In Figure 1.8 we see that the initial bearing produces a probability distribution on target location with substantial range uncertainty. The only range information that we have is that the target is no more than 6.5 nm from ownship at the time of the initial bearing. In this figure and subsequent ones we show a random sample of 500 of the 25,000 particles. The velocity distribution at this time looks like the one in Figure 1.5.

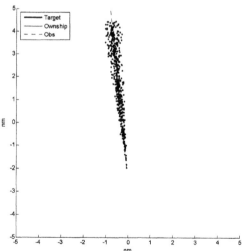

Figure 1.8   Target position distribution after the initial bearing observation

As more bearings are received, the range and bearing uncertainty are reduced somewhat as seen in Figure 1.9, which shows the position and velocity distributions after 12 minutes and 12 measurements. The target's velocity is $(4, -4)$, which is indicated by the crosshairs on the velocity distribution. Notice that the velocity distribution is substantially different from the prior. However, it is not until ownship maneuvers at 18 minutes and obtains the additional measurements through time 28 minutes that the range and velocity uncertainties are substantially reduced as shown in Figure 1.10.

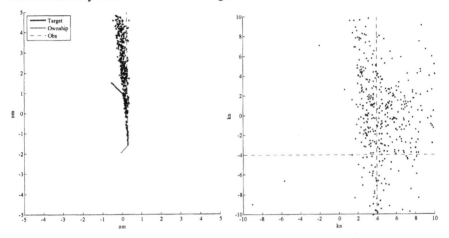

Figure 1.9   Position and velocity distributions at 12 minutes

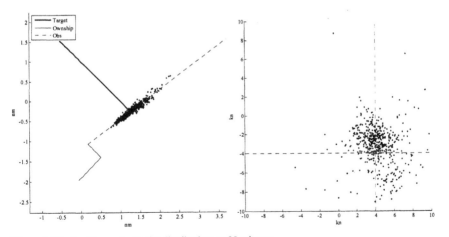

Figure 1.10   Position and velocity distributions at 28 minutes

When the target maneuvers at 36 minutes, the position and velocity uncertainty begin to increase. At 42 minutes, six minutes after the target maneuver, the position and velocity distributions shown in Figure 1.11 reflect the increased uncertainty in the position and velocity. The velocity distribution is beginning to react to the maneuver with a small number of particles located in the vicinity of the target's new velocity. Notice the "hole' in the velocity distribution around $(0,0)$ resulting from the constraint that the target speed must exceed 2 kn.

At 42 minutes, ownship maneuvers to obtain a better solution. Figure 1.12 shows the position and velocity distributions at 50 minutes. There is a large range uncertainty, but the velocity distribution now has substantial mass near the true target velocity. At the end of the example at 60 minutes, Figure 1.13 shows good localization in both position and velocity.

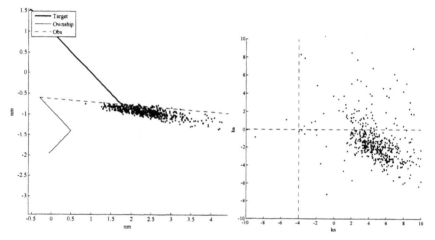

Figure 1.11   Position and velocity distributions at 42 minutes

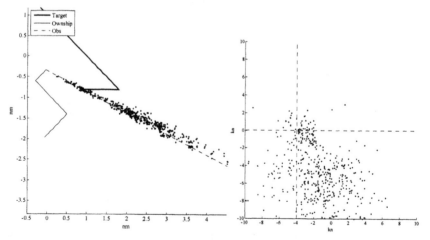

Figure 1.12   Position and velocity distributions at 50 minutes

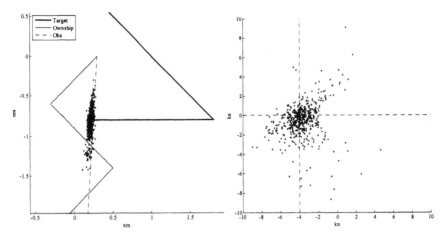

Figure 1.13   Position and velocity distributions at 60 minutes

1.3.7.1 Conclusions

This example demonstrates the power and flexibility of nonlinear tracking as implemented by a particle filter. A realistic motion model based on the tactical situation was easily implemented along with constraints on the speed of the target. The motion model allowed for target maneuvers in a natural fashion. The measurement model readily handled observations that are not linear functions of target state and allowed for the incorporation of limited range information.

At each time the posterior distribution accurately represented the uncertainty in target state. The question of the observability of the target state does not arise. Instead, the lack of range information is represented by the range uncertainty shown in the position distributions before ownship maneuvered. The filter easily handled the target maneuver, because the possibility of a maneuver is built into the motion model.

The reader is invited to compare the ease and flexibility of the particle filter approach to the numerous clever and complex approaches to performing bearings-only tracking with a Kalman filter that are presented in Chapter 6 of [2] and to recall that in the tests results presented in [2], particle filters outperformed the Kalman filters.

1.4 EXAMPLE 3: PERISCOPE DETECTION AND TRACKING

Periscope detection by shipboard radar remains an important naval problem because of the vulnerability of ships to attack by diesel-electric submarines, which are very difficult to detect acoustically when they are running on their batteries. This vulnerability has been demonstrated in numerous exercises. The U.S. Navy has responded to this threat by installing high-resolution shipboard radars on carriers to detect the periscope of a submarine making a stealthy underwater approach. During an approach, the submarine typically raises its periscope above the water in order to "look around." At these times it is vulnerable to detection by the radar.

The example in this section demonstrates the use of an LRDT as described in Chapter 7 to detect submarine periscopes (exposed for only a few seconds at a time) within a 10-nm radius of a ship without generating a large a number of false alarms.

The radar generates returns in range and bearing having a 1-ft range resolution, a 2-degree beam width, and a 5-Hz scan rate. It is assumed that a periscope will be exposed on the order of 10 seconds, although this information is not directly used by LRDT. Because the radar is mounted on a ship and is looking out to ranges of up to 10 nm, the grazing angle of the radar signal is low to the surface of the ocean. A high-resolution radar with a low grazing angle encounters significant clutter from breaking waves, which generate substantially higher returns than the mean ambient level. These high-intensity clutter spikes produce a high clutter rate and potentially a high false alarm rate (i.e. high rate of peaks in the radar return that "look like" targets).

The statistical behavior of this spiky clutter is described by a two-scale model. In this model the received radar signal is represented as a compound process where a fast speckle process is modulated by a slower process describing the scattering features in a radar cell. Over short time intervals (up to approximately 250 ms) the intensity observed in a fixed-range cell is Rayleigh-distributed. Over longer times, the mean of the Rayleigh component follows a gamma distribution whose shape parameter is a function of the radar and ocean parameters such as grazing angle, resolution cell size, frequency, look direction, and sea state.

Since the radar is looking for a submarine periscope, it is reasonable to assume that there is at most one target at any time within a 10-nm radius of the ship. The result is that we have a single (or no target) problem with a potentially large number of false alarms. To tackle this problem [20] used LRDT. The tracker employed consists of a clutter tracker and a target tracker. The clutter tracker estimates the mean clutter level (i.e., the mean of the Rayleigh distribution mentioned above) in each range cell, every 1/5 of a second using the intensity of the radar returns.

1.4.1 Target Tracker

Each beam of the radar is treated separately. The nominal periscope is up for only a short period of time, and the chance of transiting from one beam to another in this short period of time is small, especially with overlapping beams. Within each beam, the target's state space is two-dimensional: *range* and *range rate*. All quantities are measured relative to the radar. The tracker does not try to estimate motion orthogonal to the look direction because with the scales involved it is difficult to use a single radar to "triangulate" the target.

For each scan, the measurement used by this tracker is the set of observed intensities of the radar return from each range cell. The observed intensity and the estimate of the mean clutter level for a cell are used to compute a measurement likelihood ratio statistic in that cell. This statistic is the ratio of the probability of receiving the return given the target is in the cell (i.e., the periscope is exposed) to the probability of receiving the return given no target present. The two probabilities are conditioned on the estimate of the mean clutter level in the cell. This measurement likelihood ratio function is combined by pointwise multiplication with the motion-updated likelihood ratio for the target to produce the posterior likelihood ratio function over the target state space. The resulting likelihood ratio surface is then updated for motion (using the motion model described below) to form the motion-updated likelihood ratio for the next time increment.

Over the short period of exposure of a periscope, the tracker assumes a constant course and speed for the target. This produces a constant range rate for the target. Let $(r, \dot{r})$ represent a range and range-rate cell in the target state space. The likelihood ratio in cell $(r, \dot{r})$ is displaced to the cell $(r + \Delta t \dot{r}, \dot{r})$ over a time interval Δt. In this motion model, each velocity hypothesis can be treated independently.

If the radar observations are consistent with a velocity hypothesis, peaks will develop in the likelihood ratio function at position-velocity cells consistent with that hypothesis. Viewed over time, the peaks will lie along a straight line corresponding to the target's track in range versus time. Peaks that occur as noise and that do not form according to a velocity hypothesis will not be reinforced. Typically one sets a threshold level and calls a detection whenever a peak exceeds the threshold. The state at which the peak occurs is used as the estimate of the target state. This process is discussed in Chapter 7. The process of integrating the likelihood ratio functions over time according to velocity hypotheses provides much of the power of the likelihood ratio tracker methodology.

1.4.2 Example

The likelihood ratio tracker described above was applied to recorded clutter data with an injected target signal to produce the results shown below. Figure 1.14 shows the radar scan data (intensity on logarithmic scale) for 60 seconds of data and for a 1,000-ft range interval. In Figure 1.14, time increases down the page, and range increases from left to right. Each horizontal line in the figure represents a single scan, and there are 300 scans of data shown. The gray scale has been adjusted to vary from white for the minimum value to black for the maximum.

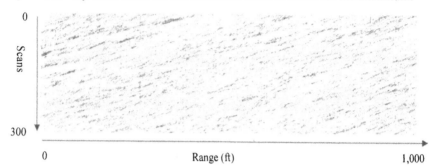

Figure 1.14   Simulated radar scan data from a single beam. High-intensity returns are shown in black with lower intensities shown on a gray scale with white equal to the lowest.

A target was injected amongst the clutter for a 10-second interval. Figure 1.14 shows the clutter problem faced by the radar in this situation. The clutter patches show up as darker areas moving toward the radar. The patches appear predominantly along the crests of long waves. It is very difficult to see the target track in this view. A single uniform threshold that was set low enough to capture a substantial number of the target returns would also let pass a large number of false alarms. It is clear that any simple thresholding scheme will either call an overwhelming number of false alarms or provide a very low detection probability.

Figure 1.15 shows the output of the likelihood ratio tracker. The axes and dimensions in Figure 1.15 are identical to those in Figure 1.14. Each horizontal line shows the logarithm of the posterior likelihood ratio function calculated through that scan. This is called the cumulative log likelihood ratio. Each dot on the line shows the cumulative log likelihood ratio in that range at that time. Darker dots indicate higher log likelihood ratios. This is a marginal log likelihood ratio obtained by adding the likelihood ratios corresponding to all velocity hypotheses in that range cell and taking the logarithm of the resulting sum. This marginal log likelihood ratio is shown in Figure 1.15 for convenience of display. The actual comparison of peak values to a threshold is performed on the likelihood ratios as a function of both position and velocity.

The results shown in Figure 1.15 are striking. The target track, which is not visible in Figure 1.14, is shown clearly in the box in Figure 1.15. It has a peak likelihood ratio value of 4×10^7 compared to the value of 7.6 for the next highest peak in Figure 1.15. If we draw the path of a target with a constant velocity (range rate) in Figure 1.15, it will appear as a straight line. The line will slant to the right if range is increasing and to the left if it is decreasing. The likelihood peaks line up very well along the target path to form an almost straight line. The peaks disappear when the periscope submerges.

An expanded view of the cumulative log likelihood ratios in the box is shown in Figure 1.16(a). Here we can see the buildup of the log likelihood ratio along the target track. Figure 1.16(b) shows the marginal log likelihood ratio in velocity (range rate) for the same time period as shown in Figure 1.15. Here we can see a clear peak at the correct range rate. This is the one corresponding to the line of peaks shown in Figure 1.15 and in Figure 1.16(a).

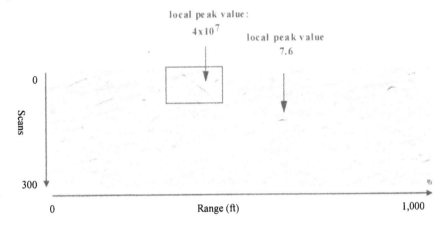

Figure 1.15 Cumulative log likelihood ratio as function of scan and range. Darker dots indicated higher log likeihood ratios.

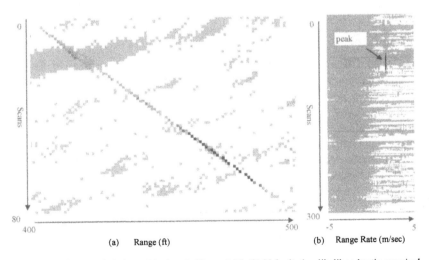

Figure 1.16 (a) Expanded view of the box in Figure 1.15. (b) Velocity log likelihood ratio marginal.

1.4.2.1 Conclusions

In this example the tracker uses unthresholded sensor output. Specifically, it does not rely on a thresholding scheme to call contacts and produce sensor measurements. As a result it can accumulate sensor information from a number of sensor responses over time, all of which may be below a threshold value, until the cumulative result crosses a threshold and allows the tracker to simultaneously call a detection and provide a track estimate.

Since the tracker has been accumulating sensor responses along many possible tracks, it has, in some sense, been performing tracking before detection. In the example shown here, the tracker implicitly considers over 30 million possible tracks to determine if the cumulative likelihood ratio for any one of them exceeds a specified threshold.

The performance of this tracker was compared to the original detector used in the Carrier periscope detection radar system. In high clutter situations the likelihood ratio tracker improved detection performance by up to 12 dB in the sense that it achieved the same detection and false alarm performance at 12-dB lower SNR for the target.

1.5 EXAMPLE 4: TRACKING MULTIPLE TARGETS

In this section we provide a simple multiple-target tracking example. There are two targets whose tracks over the time period from 0 to 2 hours are shown in Figure 1.17. Target 1 is moving to the right at 5 kn. Target 2 is moving on a 20-degree downward diagonal to the right at 5.3 kn. The target tracks cross at 1 hr, at which time the target positions coincide.

The measurements are position estimates with bivariate normal errors. Each track has been initiated with a measurement having a circular normal error; the resulting prior position distributions are shown in Figure 1.17. The circles shown are the the 2.45-σ ellipses (circles) for these distributions. The 2.45-σ ellipse has probability 0.95 of containing the target position, so we call this the 95% containment (or uncertainty) ellipse.

We use the same motion model as we did for the bearings-only tracking example in Section 1.1.2 with the exception that target speeds are constrained to be between 2 and 12 kn; the mean time between course changes is 1 hr; and $\mu_h = 20\,\mathrm{deg}$, and $\sigma_h = 10\,\mathrm{deg}$. As before, the initial distribution on target velocity has the annular form shown in Figure 1.5. The measurements are generated every 0.1 hr over a period of two hours. At each time only one measurement is generated; the measurements alternate between the two targets, first on target 1 then on target 2. (The tracker does not know this.) The values of the measurements are obtained by adding an offset drawn from the measurement error distribution to the actual target position at the time of the measurement.

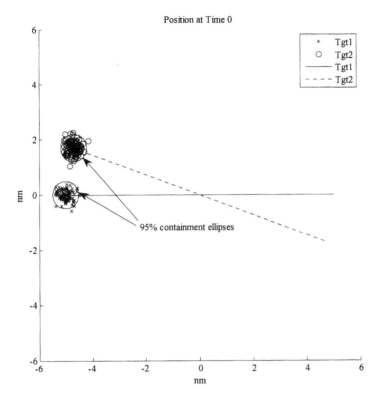

Figure 1.17 Tracks and prior position distributions for targets 1 and 2. The circles show the 95% containment ellipses for the prior distributions on the locations of the targets.

1.5.1 Soft Association

If we know there are N targets, and for each measurement we know whether it was generated by a false alarm or by a specific one of the N targets, then the multiple target problem reduces to N single target tracking problems. The difficulty is that we generally do not know the number of targets and which target (if any) generated a measurement. Chapter 4 discusses multiple target tracking and develops optimal Bayesian methods for performing such tracking. However, the methods are computationally infeasible, so approximations are necessary.

If we identify a specific target as the source of a measurement, then we say that the measurement is *associated* to that target. In multiple target tracking, one usually assumes that at each measurement time (or scan) each target generates at most one measurement and that each measurement is associated to at most one target. This is a zero-one or *hard association* assumption; either a measurement is associated with a target or it is not.

One type of multiple target approximation involves the use of *soft association* in which one can specify a probability $0 < p < 1$ that a measurement is associated with a given target. The techniques called probabilistic data association (PDA) and JPDA developed in [10] use soft association. PDA applies to the situation where there is one target, but the measurements may be generated by the target or clutter. In JPDA, there are multiple targets, and the number of targets is known. PDA and JPDA were developed for Kalman filters. In Chapter 4 we develop a version of JPDA that is appropriate for particle filters. Generalizing the JPDA methodology to particle filters allows us to perform multiple target tracking while retaining the advantages of Bayesian nonlinear tracking

As we mentioned above, the main difficulty in multiple target tracking arises from ambiguity in associating measurements to targets. An example of a measurement with an ambiguous association is shown in Figure 1.18. The elliptical uncertainty region of the measurement overlaps with two target distributions. The measurement could reasonably have been generated by either target. We now show how to compute the probability that a measurement is associated to a target and how to use this computation to perform soft association.

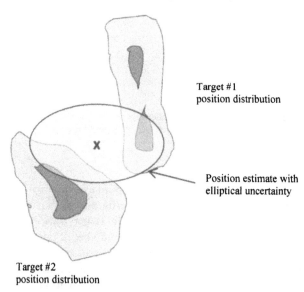

Target #1
position distribution

Position estimate with
elliptical uncertainty

Target #2
position distribution

Figure 1.18   Position measurement that may be due to either target 1 or 2.

In this example, we show the use of soft association for tracking two targets when the measurements are position estimates with bivariate normal errors. The example is a special case of JPDA where the measurements are received sequentially, one at a time and there are no false alarms. The particle filter version of JPDA developed in Chapter 4 allows for scans of data (multiple measurements received at the same time) and relaxes the assumption that the number of targets is known.

1.5.1.1 Association Probabilities

Suppose that there are M_t target tracks at the time t of a measurement and that $p_t(m, \cdot)$ is the motion-updated probability distribution on the state of the mth target for $m = 1, \ldots, M_t$. In addition to the M_t existing tracks, we add $m = 0$ for a target that has not yet been detected. The probability distribution $p_t(0, \cdot)$ for this target is often taken to be uniform over the area of interest.

Suppose that we have received a single measurement $Y_t = y_t$ at time t. Let

$$L(y_t \mid x) = \Pr\{Y_t = y_t \mid X(t) = x\}$$

be the likelihood function for the measurement given the target is located at x. The method for computing association probabilities for a measurement is derived in Chapter 4. It begins by computing association likelihoods $l(m)$ for the mth target as follows:

$$l(m) = \int L(y_t \mid x) p_t(m, x) dx \text{ for } m = 0, \ldots, M_t$$

and then computing the association probabilities

$$\alpha(m) = \frac{l(m)h(m)}{\sum_{m'=0}^{M_t} l(m')h(m')} \text{ for } m = 0, \ldots, M_t$$

where $h(m)$ is the prior probability that the measurement is associated to target m and

$$\sum_{m=0}^{M_t} h(m) = 1$$

We usually take $h(m) = 1/(M_t + 1)$.

For this example, the error in the measurement Y_t at time t has a bivariate normal distribution with mean 0 and covariance Σ_t. Thus if target position is x,

$$Y_t = x + \varepsilon_t \text{ where } \varepsilon_t \sim \mathcal{N}(0, \Sigma_t)$$

Thus the likelihood function for the measurement $Y_t = y_t$ is

$$L(y_t \mid x) = \Pr\{\varepsilon_t = y_t - x\}$$
$$= (2\pi)^{-1} |\Sigma_t|^{-\frac{1}{2}} \exp\left(-\frac{1}{2}(y_t - x)^T \Sigma_t^{-1}(y_t - x)\right) \qquad (1.7)$$

1.5.1.2 Posterior Distributions on Target State

We use a particle filter to represent the distribution of each target. Using the particle filter for target m, we compute the posterior distribution $p_t^+(m, \cdot)$ for target m given the measurement is associated to that target. We then compute the posterior $\tilde{p}_t(m, \cdot)$ as the mixture of $p_t^+(m, \cdot)$ and the motion-updated distribution $p_t(m, \cdot)$ weighted by the association probability $\alpha(m)$, namely

$$\tilde{p}_t(m, \cdot) = \alpha(m) p_t^+(m, \cdot) + (1 - \alpha(m)) p_t(m, \cdot) \qquad (1.8)$$

The influence of a measurement is felt on each track in proportion to the probability that the measurement is associated with that track.

Since we are using a particle filter to represent the target's distribution, it is easy to implement (1.8). Let

$$p_n = \text{the probability on particle } n \text{ in } p_t(m, \cdot)$$
$$p_n^+ = \text{the probability on particle } n \text{ in } p_t^+(m, \cdot)$$

Then posterior probability on particle n in $\tilde{p}_t(m, \cdot)$ is

$$\tilde{p}_n = \alpha(m) p_n^+ + (1 - \alpha(m)) p_n \qquad (1.9)$$

1.5.1.3 Tracker Output

Figures 1.19–1.26 display a random sample of 100 of the 10,000 particles for each track. Figures 1.19 and 1.20 show the position and velocity distributions for the two targets after the measurement on target 1 at 0.1 hr. The ellipse shows the 95% uncertainty ellipse for the measurement error. Since the separation between the targets is much larger than the measurement error, there is no ambiguity in the association of this measurement to target 1. Since no measurements (beyond the initializing one) have been associated with target 2, its velocity distribution retains its annular shape. However, the velocity distribution for target 1 has shifted to show a strong bias in the positive direction on the horizontal axis.

Figures 1.21 and 1.22 show the position and velocity distributions at 0.6 hr. The velocity distributions have now separated with target 1 heading to the right, and target 2 heading diagonally to the right and down.

In Figure 1.22 note the outliers in the velocity distribution for target 1, the target that did not generate a measurement at time 0.6 hr. These arise from the velocity change model where at each time period there is some probability that a particle will choose a new velocity according to the model described in Section 1.3.3.

As the targets come close to one another compared to the measurement uncertainty, the association probabilities move away from 0 and 1 as shown in Figure 1.23. As the target paths cross from time 0.7 to 1.5 the association becomes ambiguous and the association probabilities in the figure reflect that.

Figures 1.24 and 1.25 show the position and velocity distributions at time 1.0 hr when the target positions coincide. The position distributions show substantial overlap. However, the velocity distributions maintain separation. It is this separation that allows the tracker to sort out the two tracks correctly after the cross as we see in Figure 1.26, which shows the position distributions at 1.6 hr. If the targets switched paths as they crossed, the tracker would not be able to determine this. This is also true of more complicated trackers that maintain multiple association hypotheses and compute the probabilities of each hypothesis. Unless there is additional identifying information, such as feature measurements, the tracker will confuse the identities of the targets. (The multiple hypotheses tracker will retain a small probability on the correct identity.) If there are feature measurements that distinguish the targets, these can be incorporated into a likelihood function and used to correctly preserve target identity. This is discussed in Chapter 3.

1.6 SUMMARY

This chapter describes the basic tracking problem and provides examples of four of these problems. The first problem illustrates the tracking of a surface ship using a Kalman filter. The second problem, a bearings-only problem, illustrates the use of a particle filter to solve a nonlinear non-Gaussian tracking problem. The third example demonstrates the use of LRDT to perform detection and tracking in a low SNR, high clutter rate situation. The fourth example illustrates a simple multiple target tracking problem.

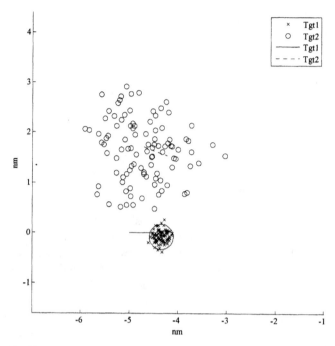

Figure 1.19   Position distributions for targets 1 and 2 at time 0.1 hr. The ellipse is the 95% measurement uncertainty ellipse.

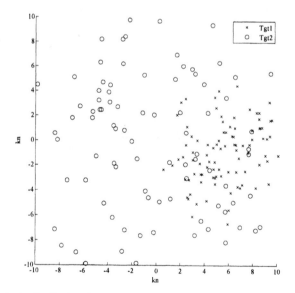

Figure 1.20   Velocity distributions for targets 1 and 2 at time 0.1 hr.

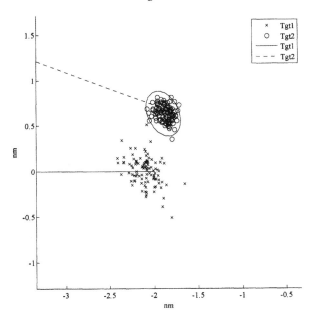

Figure 1.21 Position distributions for targets 1 and 2 at time 0.6 hr. The ellipse is the 95% measurement uncertainty ellipse.

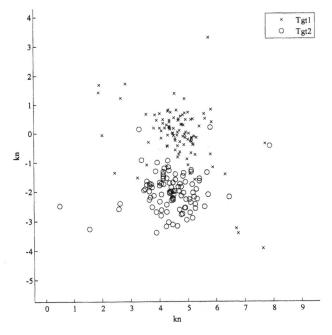

Figure 1.22 Velocity distributions for targets 1 and 2 at time 0.6 hr.

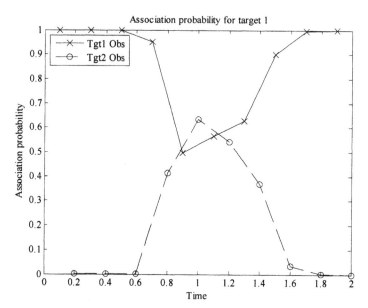

Figure 1.23   Association probabilities for target 1. The association probability for target 2 is equal to 1 minus the association probability for target 1.

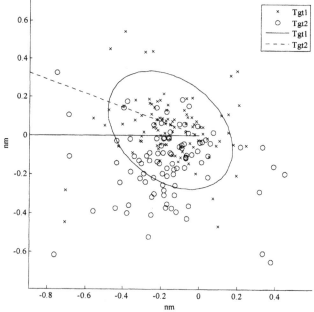

Figure 1.24   Position distributions for targets 1 and 2 at 1.0 hr. The ellipse is the 95% measurement uncertainty ellipse.

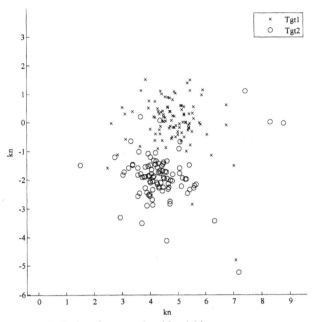

Figure 1.25   Velocity distributions for targets 1 and 2 at 1.0 hr.

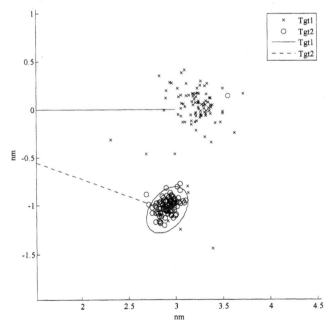

Figure 1.26   Position distributions for targets 1 and 2 at 1.6 hr. The ellipse is the 95% measurement uncertainty ellipse.

References

[1]   Kalman, R. E., "A new approach to linear filtering and prediction problems," *Journal of Basic Engineering*, Vol. 82, No. 1, 1960, pp. 35-46.

[2]   Ristic, B., S. Arulampalm, and N. Gordon, *Beyond the Kalman Filter*, Norwood: Artech House, 2004.

[3]   Bar-Shalom, Y., X. Rong Li; and T. Kirubarajan, *Estimation with Applications to Tracking and Navigation*, New York: John Wiley & Sons, 2001.

[4]   Anderson, B. D. O., and J. B. Moore, *Optimal Filtering*, Mineola, NY: Dover, 2005.

[5]   Julier, S., J. Uhlmann, and H. F. Durrant-White, "A new method for nonlinear transformation of means and covariances in filters and estimators," *IEEE Trans. Automatic Control*, Vol. 45, March 2000, pp. 477-482.

[6]   Julier, S., and J. Uhlmann, "Data fusion in nonlinear systems," in *Handbook of Multisensor Data Fusion* D. L. Hall and J. Llinas (eds.), Chapter 13, Boca Raton, FL: CRC, 2001.

[7]   Sorenson, H. W., and D. L. Alspach, "Recursive Bayesian estimation using Gaussian sums" *Automatica*, Vol. 7, 1971, pp. 465–479.

[8]   Aslpach, D. L., and H. W. Sorenson, "Nonlinear Bayesian estimation using Gaussian sum approximations," *IEEE Trans. Automatic Control*, Vol. 17, No. 4, 1972, pp. 439–448.

[9]   Blom, H. A. P., and Y. Bar-Shalom, "The interacting multiple model algorithm for systems with Markovian switching coefficients," *IEEE Trans. Automatic Control*, Vol. 33, August 1988, pp. 780-783.

[10] Bar-Shalom, Y., and T. E. Fortman, *Tracking and Data Association*, New York: Academic Press, 1988.

[11] Blackman, S., and R. Popoli., *Modern Tracking Systems*, Norwood: Artech House, 1999.

[12] Benes, V. E., "Exact finite dimensional nonlinear filters with certain diffusion nonlinear drift," *Stochastics*, Vol. 5, 1981, pp. 65-92.

[13] Daum, F. E., "Exact finite dimensional nonlinear filters," *IEEE Trans. Automatic Control*, Vol. 31, No. 7, 1986, pp. 616-622.

[14] Daum, F. E., "Beyond Kalman filters: practical design for nonlinear filters," in *Proceedings SPIE*, Vol. 2561, 1995, pp. 252-262.

[15] Sorenson, H. W., "Recursive Estimation for nonlinear dynamic systems," in *Bayesian Analysis of Time Series and Dynamic Models*, J. C. Spall (ed.), New York: Marcell Decker, 1988.

[16] Sorenson, H. W., "On the development of practical nonlinear filters," *Information Sciences*, Vol. 7, 1974, pp. 253-270.

[17] Stone, L. D., C. A. Barlow, and T. L. Corwin, *Bayesian Multiple Target Tracking*, Norwood: Artech House, 1999.

[18] Doucet, A., N. de Freitas, and N. Gordon (eds.), *Sequential Monte Carlo Methods in Practice*, New York: Springer, 2001

[19] Kahn, H., *Use of Different Monte Carlo Sampling Techniques*, The Rand Corporation (P-766), 1955 (http://www.rand.org/pubs/authors/k/kahn_herman.html).

[20] Finn, M. V., et al., *Uncluttered Tactical Picture*, Metron Inc Report to Office of Naval Research, 17 April 1998 (http://lib.stat.cmu.edu/general/bmtt.pdf).

Chapter 2

Bayesian Inference and Likelihood Functions

The objective of tracking is to estimate the number and state of targets in a region of interest. The term *state* is used to indicate a vector of quantities that characterizes the object being tracked in a way that is discussed in Chapter 3. By definition, tracking is an estimation problem, in fact, a statistical estimation problem. It is our belief that this problem is best viewed as a Bayesian inference problem.

In Section 2.1, we present our rationale for the choice of Bayesian inference as the method for dealing with these problems. Readers who are interested in the applications of Bayesian statistical methods but not their philosophical underpinnings can skip Section 2.1 and begin with Section 2.2. The authors have found that Bayesian inference provides a principled and effective method of approaching most inference problems, especially nonstandard ones. McGrayne [1] provides a history of the development, use, and acceptance of Bayesian inference. The range of applications is surprisingly wide and powerful.

This rationale for the use of Bayesian inference is provided in terms of a philosophical discussion comparing classical and Bayesian statistical estimation philosophies. The reader may wonder why we have limited our consideration to these two. Both of these methods are mature and have withstood the test of time. Other methods of inference such as Dempster-Shafer theory [2], fuzzy logic [3], and conditional event algebras [4] are interesting and even promising, but they have not, in our view, achieved the completeness of either Bayesian or classical statistical estimation. For example, in Dempster-Shafer theory, the state space must be discrete and basic computations explode exponentially as the number of points in the state space increases. In addition Dempster-Shafer lacks a well-defined notion of expected value. Chapters 1 and 9 of [5] detail some of the shortcomings of these alternate methods of inference.

Bayesian inference and Bayesian tracking require the use of likelihood functions. In Sections 2.2 and 2.3 we define likelihood functions and give examples of ones used in tracking problems. Likelihood functions are the common currency of information in a Bayesian tracker. Bayesian inference and the use of likelihood functions allow one to incorporate disparate information into an estimate in a way that other methods do not.

2.1 THE CASE FOR BAYESIAN INFERENCE

As an estimation problem, tracking is challenging since the typical problem is to estimate a parameter that is changing as one is receiving data. To make matters worse, the data can be remarkably uninformative and obtained from a number of different sources.

A reading of the classical statistical literature on parametric estimation would lead one to methods for computing estimators such as the maximum-likelihood estimator. A good deal of space would be dedicated to describing the estimator under ideal circumstances (such as repeated independent sampling) and considering its asymptotic properties, sufficiency, consistency, and efficiency. Multiple target tracking involves circumstances that are far from ideal. In many tracking problems, special assumptions such as normality or linear relationships between measurements and target state fail to hold. Tracking requires methods that work not only in the ideal cases of normality and repeated sampling but also in the cases where sample size is small and special assumptions do not hold.

Reference [6] provides an introduction to classical mathematical statistics. References [7, 8] provide more breadth and mathematical depth. The three volumes comprising [9] contain an encyclopedia of classical statistical methods.

2.1.1 Frequentist Point of View

When estimation is stripped of special assumptions, one is left with foundations. A fundamental debate that occurs at the root of inference is over the criteria for judging the quality of a statistical procedure. On one hand, classical statisticians argue that an estimator should be judged by its long-run convergence properties over many experiments. For example, an estimator is judged effective if it can be demonstrated that it converges to the true parameter value regardless of the value of that parameter. This is called the frequentist view. This view is so named since, under this regimen, statistical procedures are judged by their long-run frequency characteristics. On the face of it, such an approach sounds very reasonable. However, such methods necessarily involve assumptions about the relative frequencies of events that are highly unlikely to hold in actual experiments. A question then occurs: Why should the statistician permit probabilities associated with rare events that did not occur to affect inference performed on data that did occur?

The frequentist point of view is reasonable when one is conducting a large number of independent, repeated trials under identical conditions such as flipping a coin. In this case it makes sense to judge estimators based on their long-term properties. This is not the situation with tracking. We are not able to repeat a tracking problem 1,000 times under identical conditions. Instead we must use methods that work for estimation problems that are one-of-a-kind since in most cases, each instance of a tracking problem will be unique and not repeatable.

2.1.2 Conditionalist Point of View

The above question gives rise to an alternative point of view, namely that inference should be based solely upon the events that actually occur in an experiment. This is not to say that other considerations may not be important in drawing conclusions from the experiment (such as prior information about the unknowns). However, the only *events* that will affect the conclusions will be the ones *actually* observed. This is known as the Conditionality Principle which forms the basis of the conditionalist point of view.

The concept common to the frequentist and conditionalist points of view is the likelihood function. This function is defined in Section 2.2.1. It expresses the probability of the observed events as a function of the unknown parameter (e.g., target state). As such, it is the logical link between the observed events and the unknown. The two points of view discussed above differ in the way each uses the likelihood function. The frequentist uses it to judge the average properties of an estimator or decision statistic before an experiment occurs. Such properties guide his selection of a procedure. To do this, he must analyze statistical procedures with respect to all events that might occur in the experiment. The conditionalist focuses on the events that actually occurred in the experiment. He is interested in the evaluation of the likelihood function just for the events that have actually occurred.

2.1.2.1 Likelihood Principle

In this book we adopt a specific form of the conditionality principle called the likelihood principle. The likelihood principle states that the likelihood function evaluated at the observed events is a complete summary of the information in the observations. Other exogenous information may also be used, but the information in the observations is fully represented by the likelihood of the observed events. In particular the likelihood of events not observed is irrelevant. This is the statement that distinguishes the frequentist from the conditionalist, since the frequentist needs the likelihood of unobserved events to compute the long-run properties of proposed estimators.

2.1.3 Bayesian Point of View

A Bayesian may be considered to be an individual who has adopted the likelihood principle, but who argues that it is not sufficient. The likelihood function links the events observed in the current experiment to the unknowns. As such, it represents evidence derived from the current data in favor of certain values of the unknown parameter.

A Bayesian would argue that there is always prior information as well. That is, there is always information about the unknown parameter available from sources independent of the data. That knowledge can simply be common sense: Ships do not travel on land; trucks normally stay on roads; tanks do not traverse terrain with steep inclines; and ships do not travel at 100 knots. This is all prior information in the sense that it is known prior to any observations. Furthermore, it is information in the same sense as the data to be collected; properly processed and employed it delimits the possible state of the unknown.

2.1.3.1 Prior Distributions

Prior information enters Bayesian inference through prior distributions. Prior distributions are probability distributions that summarize the information about the unknown parameter at a given time *prior* to obtaining further information from observations.

In many cases one may have objective data that can be used to construct a prior distribution about the unknown parameter. For example, when testing a person for a disease one can construct a prior probability of disease from data on the relative frequency of the disease in the population being tested. Other cases may involve estimating the value of a one-of-a-kind parameter such as the location of an aircraft wreck on the ocean bottom. One cannot run repeated experiments to determine the prior on this location.

In the Bayesian point of view, the prior distribution on the parameter of interest represents and quantifies our knowledge or uncertainty about the value of the parameter. The fact that a probability distribution is assigned to an unknown parameter does not mean that it is considered to be random, just that there is uncertainty in its value. To construct a prior it may be necessary to use subjective as well as objective information. Chapter 3 in [10] discusses a number of ways, objective and subjective, of constructing a prior. Chapter 6 of [11] provides a set of conditions that guarantee the existence of a subjective probability distribution consistent with a statistician's estimate of the relative probabilities of various values of the parameter.

Classical statisticians view parameters as fixed, even if unknown, and reject the idea of any "distribution" for them. They also argue that even if one accepts the concept of such a distribution, its exact nature is almost inevitably subjective, possibly based on some vague degree-of-belief concept rather than the "scientifically-sound" concept of relative frequency in repeated experiments. Alternatively, Bayesian statisticians are willing to use prior distributions for unknown parameters, even if there is considerable subjectivity involved. Bayesian statistical methods are presented in [10–13].

2.1.3.2 Bayes' Theorem

Prior distributions are essential to the use of Bayes' theorem. This theorem, given in a paper by Bayes [14] (see [15] for a reprint of Bayes' original essay), is generally cited as the beginning of this methodology. However, it is clear from [1] that Laplace deserves at least as much credit for what we call Bayes' theorem as Bayes himself. Laplace independently discovered Bayes' theorem, extended it to nonuniform prior distributions, and applied it to many important scientific problems. In fact the form in which we write Bayes' theorem today (see (2.2)) is due to Laplace. Bayes' theorem is a powerful inference tool. It is surprising that such a powerful tool is a simple consequence of the definition of conditional probability.

The examination of Bayes' theorem provided in Section 2.2.2 shows that it is a method of resolving current evidence with prior information. The prior distribution on the space of the unknown parameter (i.e., the parameter space) is the first evidential weighting function. The likelihood function deriving from the current observations, when viewed as a function on the parameter space, then serves as a second evidential weighting function on that space. Typically, these two functions will be different. Bayes' Theorem says that the way to resolve this difference is to multiply these two evidential functions on the parameter space and then to renormalize to yield a new probability distribution called the posterior. One can think of this new distribution as the prior for the next round of observations.

2.1.3.3 Benefits of Bayesian Inference

To a Bayesian, the posterior distribution on the parameter space is the natural answer to any inference question. To others, more specific answers are required. However, a Bayesian claims that all natural answers may be derived from the posterior. For example, the Bayesian equivalent of a "good" frequentist point estimator is the mean of the posterior distribution. In fact the mean is the optimal estimator if one seeks to minimize the expected squared error between the estimator and the actual value. More generally, if a decision procedure is needed to minimize some measure of loss associated with being incorrect, the decision procedure is judged by computing the expected loss using the posterior distribution. If one needs to establish a set of parameter values likely to contain the true value with a specified level of assurance, a Bayesian can use the posterior to find the smallest set of parameter values with the stipulated posterior probability. From the Bayesian point of view, the posterior is the starting point for addressing all well-posed statistical questions. This approach has the benefit of simplifying and unifying the practice of statistics.

Another benefit of the Bayesian point of view is that when observation errors are mutually independent, Bayes' theorem is naturally recursive. For example, consider two observations. They may represent observations obtained from repeated and identical experiments, or they may not. Bayes' theorem requires that one compose a likelihood function on the parameter space that expresses the likelihood of both observations. It is this joint likelihood that one multiplies by the prior to yield the posterior. Now consider a recursive method of analyzing the data from these two observations. In this method, one computes the likelihood function for the first observation and combines it with the prior according to Bayes' theorem to produce the posterior for the first observation. Assuming independence of the errors in these two observations, one computes the likelihood function for the second observation and combines it with the posterior from the first observation to obtain a posterior for both observations. In this step, one has used the posterior from the first observation as the prior for the second. We show in Section 2.2.2 that as long as the parameter space remains the same for the two observations and the observational errors are independent, the recursive procedure yields the correct Bayesian posterior distribution.

The recursive nature of Bayes' theorem makes it particularly appropriate for the type of estimation that arises in tracking problems. The discussion above concerns the problem of estimating the value of a fixed parameter, one whose value is unknown but that does not change during observation. Recursive Bayesian processing as described above (with a little generalization) may also be employed to estimate the value of a parameter that changes with time, such as target state. Additional arguments for Bayesian inference may be found in [10].

There are some drawbacks to Bayesian inference. It requires the use of a prior distribution whose generation may necessitate extensive assumptions. If there is a paucity of data, the conclusions may be driven by the assumptions used to form the prior. Finally, in some cases, Bayesian inference involves the use of subjective probabilities, which creates a sense of unease in some people.

2.2 THE LIKELIHOOD FUNCTION AND BAYES' THEOREM

This section defines the notion of likelihood function that we will use in this book and discusses Bayes' theorem.

2.2.1 The Likelihood Function

Likelihood functions replace and generalize the notion of measurement used in linear Gaussian trackers. For this reason, likelihood functions play a key role in Bayesian tracking.

Let X be a random variable whose values fall in the state space S. Let us suppose that information about the value of X is obtained through observations or measurements Y whose values fall in the space M and depend on X. Specifically we assume Y is a random variable whose distribution depends on X. The likelihood function expresses the relationship between X and Y.

Definition 2.1. The *likelihood function* L for the random variable X and measurement $Y = y$ is defined to be

$$L(y \mid x) = \Pr\{Y = y \mid X = x\}$$

where $L(\cdot \mid x)$ is a probability density function on M for each $x \in S$. In some cases the probability density function may be subjective. When we have a function of two (or more) variables such as $L(y \mid x)$, the notation $L(\cdot \mid x)$ designates the function of one variable obtained by fixing the value of x and letting y vary.

Each measurement will be represented by a different Y and may have a different measurement space M.

We use the notation $\Pr$ to mean either probability or probability density as appropriate and the term probability density function to include discrete as well as continuous density functions. Specifically, there is a measure on the space M, and for each $x \in S$, $L(\cdot \mid x)$ is a probability density with respect to that measure. If M is discrete, then $L(y \mid x)$ is a probability for each $y \in M$.

Once a measurement is made, the data $Y = y$ are fixed. Generally, we will consider likelihood functions $L(y \mid \cdot)$ where y is fixed and x varies over the space S. When viewed as a function of the variable to estimated, $L(y \mid \cdot)$ is not a probability density function on S but is instead a likelihood. The utility and power of this point of view is that it allows us to combine information from disparate sources having different measurement spaces on the common state space S.

R. A. Fisher was the first to use the term likelihood. His reason for the name likelihood function is that if the measurement is $Y = y$ and $L(y \mid x_1) > L(y \mid x_2)$, then x_1 is more likely to be the true value of X than x_2 (assuming that x_1 and x_2 have equal prior probabilities).

2.2.2 Bayes' Theorem

Suppose that Y is a measurement whose probability distribution depends on the value of a random variable X so that we can compute the likelihood function

$$L(y \mid x) = \Pr\{Y = y \mid X = x\} \text{ for } x \in S \tag{2.1}$$

Let p be the probability density function for X. The conditional probability density for X given the measurement $Y = y$ is

$$p(x \mid y) = \frac{\Pr\{Y = y \mid X = x\} p(x)}{\Pr\{Y = y\}} = \frac{L(y \mid x) p(x)}{\int L(y \mid x) p(x) dx} \tag{2.2}$$

where the integration in the denominator becomes a summation if the probability distribution for X is discrete. The statement in (2.2) is usually referred to as Bayes' theorem. The distribution given by p is called the *prior* distribution and the distribution given by $p(\cdot \mid y)$ is called the *posterior* distribution for X given the measurement $Y = y$.

From (2.2) we see that

$$p(x \mid y) = C^{-1} L(y \mid x) p(x) \tag{2.3}$$

where the constant C equals the denominator in (2.2). Note C is constant with respect to x but not y. Here we are treating the measurement $Y = y$ as data and fixed but the state or parameter value $X = x$ as unknown.

The posterior is proportional to the likelihood times the prior. Equation (2.3) shows clearly that it is through the likelihood function that the measurements (data) modify the information represented by the prior distribution on X. The likelihood function can be regarded as representing the information about X contained in the measurement Y. In this book, likelihood functions will be the common currency by which information is represented and incorporated into the tracking process.

2.2.3 Sequential Nature of Bayes' Theorem

Bayes' theorem in (2.2) is appealing because it provides a mathematical formulation of a method by which previous knowledge may be combined with new knowledge. Indeed, Bayes' theorem allows us to continually update information about a parameter X as more measurements are taken.

Suppose we have an initial measurement $Y_1 = y_1$ with likelihood function L_1. Then Bayes' theorem yields

$$p(x \mid y_1) \propto L_1(y_1 \mid x) p(x) \qquad (2.4)$$

Suppose we have a second measurement $Y_2 = y_2$ with likelihood function L_2 where Y_1 and Y_2 are independent given $X = x$. Then

$$p(x \mid y_1, y_2) \propto L(y_1, y_2 \mid x) = L_2(y_2 \mid x) L_1(y_1 \mid x) p(x)$$
$$\propto_2 L_2(y_2 \mid x) p(x \mid y_1) \qquad (2.5)$$

The expression in the second line in (2.5) has precisely the same form as the one in (2.4) except that the posterior distribution for X given $Y_1 = y_1$ plays the role of the prior distribution for the second sample. Obviously this process can be repeated any number of times. In particular, if we have n independent measurements, the posterior distribution can, if desired, be recalculated after each new measurement, so that at the mth stage the likelihood associated with the mth measurement is combined with the posterior distribution of X after $m-1$ measurements to give the new posterior distribution.

Thus Bayes' theorem describes in a fundamental way the process of learning from experience and shows how knowledge about the state of nature represented by the posterior distribution on X is continually modified as new data become available.

2.3 EXAMPLES OF LIKELIHOOD FUNCTIONS

We now present examples of likelihood functions for some typical sensors used in tracking.

2.3.1 A Gaussian Contact Model

One of the most common forms of information available on the state of a target is a Gaussian contact. This is a measurement that is a linear function of the target state perturbed by additive Gaussian error ε. We will use the notation $\varepsilon \sim \mathcal{N}(\mu, \Sigma)$ to mean that ε has a Gaussian distribution with mean μ and covariance Σ.

Radars often provide measurements in the form of Gaussian contacts. To model this measurement, we let X be the unknown target state. We assume that X is an l-dimensional real column vector and that the measurement Y is an r-dimensional column vector. If z is a vector, we let z^T denote the transpose of that vector. Usually vectors will be column vectors and their transposes row vectors. The measurement Y is modeled as

$$Y = HX + \varepsilon \qquad (2.6)$$

where H is an r-by-l matrix of real numbers called the measurement matrix and $\varepsilon \sim \mathcal{N}(0, R)$ where R is an r-dimensional measurement-error covariance matrix. Equation (2.6) is often called the measurement equation in Kalman filtering terminology. Since ε is Gaussian, Y is also Gaussian. From (2.6), we see that likelihood function is given by

$$L_G(y|x) = \Pr\{Y = y \,|\, X = x\} = \Pr\{\varepsilon = y - Hx\}$$

$$= (2\pi)^{-r/2} |\Sigma|^{-1/2} \exp\left(-\frac{1}{2}(y - Hx)^T \Sigma^{-1}(y - Hx)\right)$$

As an example, suppose there is a radar sensor that produces contacts that are estimates of target position. Specifically we take $l = 2$, so that X is a two-dimensional vector and $H = \mathbf{I}$, where $\mathbf{I}$ is the 2×2 identity matrix. Let $\Sigma = \sigma^2 \mathbf{I}$ where σ is a scalar. Suppose that the radar produces a contact $Y = (0,0)^T$ with $\sigma^2 = 2$ nm^2. Then $L_G(y|\cdot)$ is a Gaussian density function with mean $(0,0)^T$. Figure 2.1 shows the likelihood function for the Gaussian contact $Y = (0,0)^T$ when $\sigma^2 = 2$ nm^2.

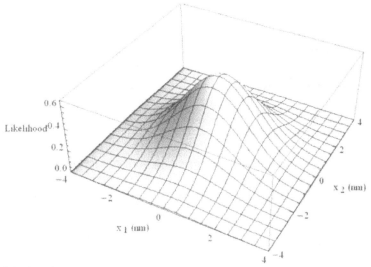

Figure 2.1   Gaussian contact likelihood function.

2.3.2 A Gaussian Bearing Error Model

A common measurement available to a tracking system is the bearing to a target. For example, the acoustic sensors on the submarines in the bearings-only tracking example in Section 1.3 produce bearing measurements. The measurements are sometimes made with beams formed from a linear array of omnidirectional sensors. In such a circumstance the bearing reported is either the direction of the center of the beam with the maximum response; or if several beams have responses close to the maximum, an interpolation of their respective beam centers. In any case, the system estimates the direction to the object under observation with error. The net result is a measurement that is a function of the state of the target with an associated likelihood function.

In this example, we model the error in the bearing measurement as a Gaussian random variable. This is appropriate as long as small errors in measurement are anticipated. In this example we assume that the target state space is two-dimensional with coordinates $x = (x_1, x_2)$.

Let $\alpha = (\alpha_1, \alpha_2)$ be the location of the sensor making the bearing measurement. Let σ_d be the standard deviation of the bearing measurement error in degrees. Let θ be the measured bearing to the target in degrees. Under our assumption that the difference between the true bearing to the target and the measured bearing θ is Gaussian, the likelihood function for the bearing measurement may be written as

$$L_B(\theta \mid x) = \left(2\pi\sigma_d^2\right)^{-\frac{1}{2}} \exp\left(-\frac{1}{2\sigma_d^2}(\theta - b(\alpha, x))^2\right)$$

where $b(\alpha, x)$ is the bearing in degrees from the sensor location α to point x.

Figure 2.2 depicts the likelihood function for a bearing measurement with a Gaussian bearing error. The acoustic center of the sensor is located at the point $(7\,\text{nm}, -7\,\text{nm})$ from the origin; the bearing measurement is 132 degrees measured counterclockwise from the positive x_1 axis; and the standard deviation of the bearing error is 5 degrees.

Notice that the likelihood function $L_B(\theta \mid \cdot)$ is not a probability density function on the (x_1, x_2) plane because it does not integrate to 1. In fact, it integrates to infinity. Most likelihood functions are not probability densities on the target state space. When $H = I$ in (2.6), the Gaussian contact likelihood function $L_G(y \mid \cdot)$ is a probability density on the target state space S. However, in general $H \neq I$ and $L_G(y \mid \cdot)$ does not integrate to 1.

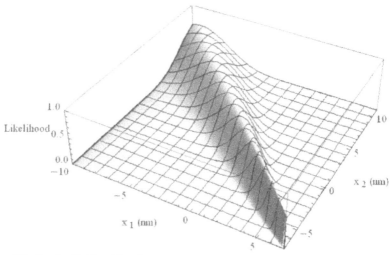

Figure 2.2   Bearing likelihood function.

2.3.3 Combining Bearing and Contact Measurements

Bearing estimates formed from linear arrays of omnidirectional sensors have an ambiguity about the axis of the array. Suppose we have a linear array of acoustic sensors being towed under the ocean surface. The sensor measurements used to construct a bearing estimate of 132 degrees are also completely consistent with the estimate that is the mirror bearing about the array axis. Figure 2.3 shows the likelihood function with the mirror bearing included. (A bearing estimate from a line array is really a three-dimensional cone whose axis coincides with the array axis. For simplicity we consider only the intersection of this cone with the plane parallel to the surface of the ocean and passing through the array axis.)

Suppose that at the time corresponding to this bearing report, we obtain the radar measurement represented by the Gaussian contact in Figure 2.1. This could be a surface ship simultaneously detected by the radar and the underwater acoustic array. In Figure 2.4, we show the array axis, the bearing of 132 degrees (solid line), and the mirror bearing of -132 degrees (dashed line). This figure also shows the location of the Gaussian contact in Figure 2.1 and its 2-σ circle of uncertainty. Since $\sigma^2 = 2$ for this contact, the circle in Figure 2.4 has a radius equal to $2\sqrt{2}$.

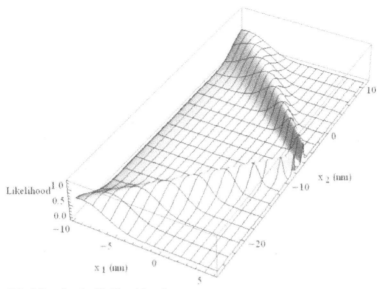

Figure 2.3 Mirror bearing likelihood function.

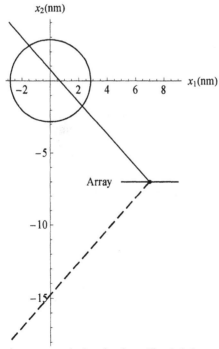

Figure 2.4 Plot of Gaussian contact and mirror bearings. The circle is centered on the Gaussian contact and has radius $2\sigma = 2\sqrt{2}$.

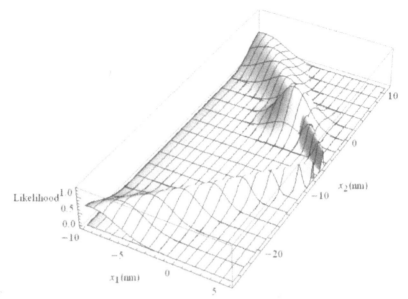

Figure 2.5 Superposition of the Gaussian contact and mirror bearing likelihood functions. Note, this is not the correct way to combine likelihood functions.

Figure 2.5 shows the likelihood function for the Gaussian contact (Figure 2.1) and the mirror bearing measurement (Figure 2.3) superimposed on one another. However, this is not the correct way to combine these likelihood functions.

From Section 2.2.2, we know that the correct way to combine two simultaneous observations with independent measurement errors is to (pointwise) multiply the likelihood functions representing the observations. Figure 2.6 shows the combined likelihood function, which is obtained by pointwise multiplication of the Gaussian contact and mirror-bearing likelihood functions. Notice that combining the mirror-bearing likelihood with the contact likelihood has "eliminated" the mirror bearing. By comparing the combined likelihood function with the contact likelihood function in Figure 2.1, we see the bearing information has reduced the uncertainty that was present in the contact information, and it has modified the mean or most likely location for the target from $(0,0)$ to a point in the first quadrant. The combined likelihood is approximately, but not exactly, a Gaussian likelihood function.

In this example we have combined information from two different measurement spaces, position space and bearing space. This is possible because a likelihood function converts a measurement to a function on the common target state space. This illustrates why likelihood functions are the common currency of information in a Bayesian tracker.

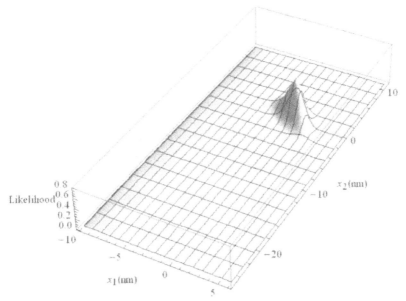

Figure 2.6   Combined Gaussian contact and mirror-bearing likelihood.

2.3.4 Negative Information

In this section we consider the likelihood for the event "search with no detection." This is as much an observation as a physical measurement. As such, it can be processed just as a physical measurement can be processed, albeit with an impact on inference reflecting its typically poorer localization content than that provided by a detection. Processing of negative information requires a model for the probability of detection. This is because the likelihood function for "negative information" or the event "no detection" is simply the probability of no detection given target location. Typically, the observation "no detection" is associated with search conducted between well-defined time limits, t_1 and t_2. For simplicity in presenting this example, we assume that the tracker is operating with discrete time increments and that the probability of failing to detect the target conditioned on target location is independent from one time increment to the next.

For this example we consider a passive acoustic underwater sensor. It has the following detection model. We assume that the "strength" in decibels (dB) of the acoustic signal radiated by the target is L_s and that during propagation of that signal between the target state x and the sensor location α, the signal strength is attenuated by an amount $P_L(\alpha, x)$ (measured in decibels) so that

$$\text{expected received signal level} = L_s - P_L(\alpha, x)$$

The actual received signal level at the sensor is perturbed by a fluctuation term ε, (measured in decibels) so that

$$\text{received signal level} = L_s - P_L(\alpha, x) + \varepsilon$$

Detection depends on how much stronger the received signal is than the background noise. Let A_N be the level (in decibels) of the ambient noise at the sensor. Our detection model assumes that when the received signal level minus the ambient noise is above a threshold D_T a detection is called, and when it is below D_T no detection is called. Define

$$S_E(x) = L_s - P_L(\alpha, x) - A_N - D_T \quad \text{for } x \in S$$

to be the *mean signal excess* at the sensor given the target is at x. $S_E(x)$ is the expected amount (in decibels) by which the signal level minus ambient noise at the sensor exceeds the detection threshold D_T given the target is at x. During a time increment, the probability of detecting a target located at x (and presumed stationary during the time increment) with a sensor located at α is given by

$$
\begin{aligned}
P_d(\alpha, x) &= \mathbf{Pr}\{L_s - P_L(\alpha, x) - A_N + \varepsilon \geq D_T\} \\
&= \mathbf{Pr}\{\varepsilon > D_T - L_s + P_L(\alpha, x) + A_N\} \\
&= \mathbf{Pr}\{\varepsilon > -S_E(x)\}
\end{aligned}
$$

For this example we assume that ε is Gaussian with mean 0 and standard deviation σ dB. The fluctuations in ambient underwater noise are often Gaussian distributed in decibels. Thus,

$$
\begin{aligned}
P_d(\alpha, x) &= \int_{-S_E(x)}^{\infty} \frac{1}{\sigma\sqrt{2\pi}} \exp\left(-\frac{z^2}{2\sigma^2}\right) dz \\
&= 1 - \int_{-\infty}^{-S_E(x)} \frac{1}{\sigma\sqrt{2\pi}} \exp\left(-\frac{z^2}{2\sigma^2}\right) dz
\end{aligned}
$$

Let the observation Y take on the value 0 for no detection and 1 for detection. Then the likelihood function for no detection ($Y = 0$) is given by

$$L(0|x) = 1 - P_d(\alpha, x)$$

$$= \int_{-\infty}^{-S_E(x)} \frac{1}{\sigma\sqrt{2\pi}} \exp\left(-\frac{z^2}{2\sigma^2}\right) dz \quad \text{for } x \in S$$

Let us consider an example where we have a sensor located at $(0,0)$ in the plane near the surface of the ocean. Suppose that

$$L_S = 60 \text{ dB}, \ A_N = 5 \text{ dB}, \ D_T = 5 \text{ dB}, \ \sigma = 8 \text{ dB}$$

and that the propagation loss function is given in Figure 2.7 as function of range from the sensor. (The acoustic community normally plots propagation loss so that the positive decibel direction is pointing down rather than up as shown in the Figure 2.7.) The propagation loss curve shown in this plot is typical of deep ocean areas where the change in water temperature with depth causes the speed of sound through the water to vary with depth. This variation produces a refraction of the acoustic rays that results in convergence zones (areas of low propagation loss) roughly every 30 nm or so as shown in Figure 2.7.

Note that $L_S - A_N - D_T = 50$. This level is shown in Figure 2.7 as a horizontal line. Whenever the propagation loss is above this level, the probability of detection is less than 0.5. When it is below 50 dB, the probability of detection is greater than 0.5. If L_S increases, this line moves upward and less of the propagation loss curve is above this line. This produces more ranges at which the probability of detection is greater than 0.5.

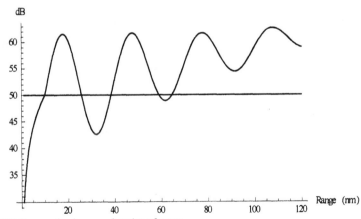

Figure 2.7 Propagation loss as a function of range.

Figure 2.8 shows the likelihood function for no detection by an omnidirectional sensor located at $(0,0)$. We have shown this likelihood function only in the first quadrant. It is symmetric about $(0,0)$. Notice that this likelihood function cannot be well approximated by a Gaussian density function. If the source level L_s increases, then the negative information likelihood function moves toward 0. This means it becomes less likely that a target is present and undetected.

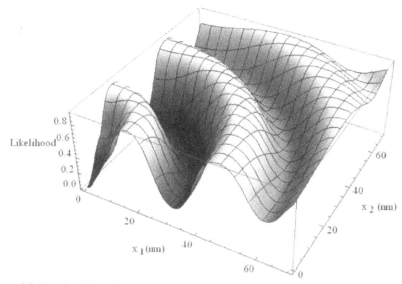

Figure 2.8 Negative information likelihood function.

2.3.5 Positive Information

Suppose that an acoustic sensor, with the detection capabilities described above, is located at $(70 \text{ nm}, 0 \text{ nm})$ from the origin and has produced a detection. Figure 2.9 shows the likelihood function for this observation. It is equal to one minus the negative information likelihood function. This shows the information in the detection part of the observation. In addition, suppose that there is a bearing measurement of 130 degrees from this same sensor (measured counterclockwise from the x_1 axis) with a standard deviation of 15 degrees. Figure 2.10 shows the likelihood function for this observation, and Figure 2.11 shows the likelihood function that results from combining these two observations. Note again that the likelihood functions have multiple modes and are not well approximated by a Gaussian density.

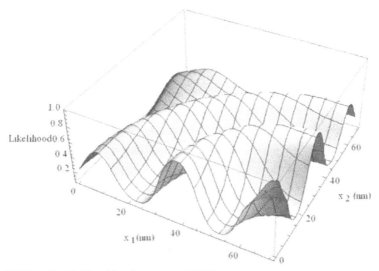

Figure 2.9 Detection likelihood function: sensor at (70,0).

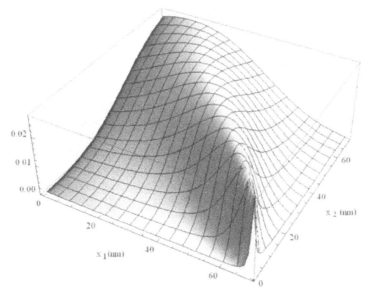

Figure 2.10 Bearing likelihood function for a sensor at (70 nm, 0 nm). The vertical axis gives the likelihood value of the function.

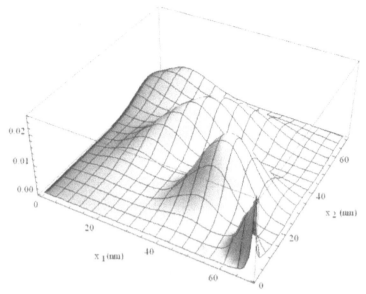

Figure 2.11 Combined bearing and detection likelihood function for a sensor at (70 nm, 0 nm). The vertical axis gives the likelihood value of the function.

2.3.6 Radar and Infrared Detection

Radar and infrared (IR) sensors are often used on unmanned surveillance aircraft to detect and track targets such as trucks or other vehicles. Typically a radar sensor has high resolution in range but lower resolution in bearing. An IR sensor often has high bearing resolution but provides very limited range information. In this example we present notional likelihood functions for colocated radar and infrared sensors and show how to combine measurements from these two sensors. The assumed error characteristics of these notional sensors are not meant to reflect the actual error characteristics but are used to illustrate the concept of how likelihood functions can be used to combine measurements from these two different types of sensors.

Suppose that the radar and infrared sensors are both located at $(0,0)$ and that the radar obtains a contact with range (gate) between 20 and 21 nm and bearing of 30 degrees. The error in the bearing is assumed to be Gaussian with mean 0 and standard deviation 2 degrees. Simultaneously, the infrared sensor detects the same target and reports a bearing of 31 degrees. The measurement error for this bearing is Gaussian with mean 0 and standard deviation 0.5 degrees.

The likelihood functions for the radar and infrared measurements are shown in Figures 2.12 and 2.13. The combined likelihood function is shown in Figure 2.14. Observe that combining the likelihood functions produces a likelihood function with a sharp peak in both range and bearing.

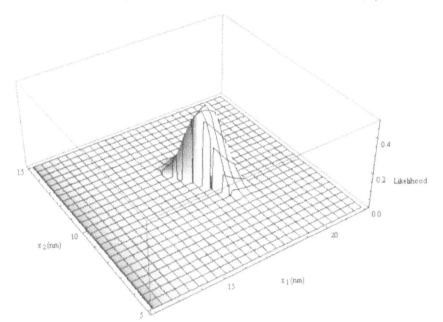

Figure 2.12 Radar likelihood function.

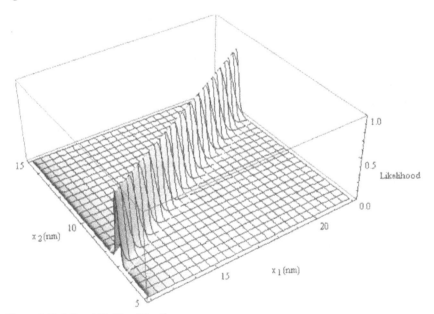

Figure 2.13 Infrared likelihood function.

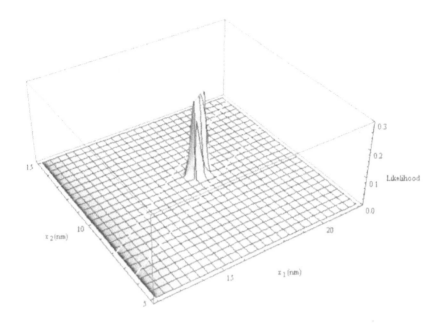

Figure 2.14 Combined radar and infrared likelihood function.

2.3.7   A Signal-Plus-Noise Model

Most often the information supplied to a tracking system is in the form of
estimates of the target's state, such as position. These estimates are often obtained
by applying a threshold to the sensor responses and performing signal processing
on the responses that cross the threshold to obtain the estimate. This example
demonstrates that by using likelihood functions and physical models one can
transform basic, unthresholded sensor measurements, such as signal plus noise,
into estimates of target state.

Tracking can be performed with simple sensors that have little or no signal
processing capability. For example, unattended ground sensors (UGSs) are placed
in an area to provide surveillance on vehicles in that area over long periods of
time. These sensors are typically battery-operated and perform a minimum of
signal processing within the sensor itself to conserve battery life. Measurements
are sent to a central station for processing into detections and tracks.

Let us consider a simplified one-dimensional example involving a stationary
target. (The more realistic problem is two- or three-dimensional with a moving
target.) Acoustic sensors are located at specified positions along a line. At a
given time the sensors report amplitude (in decibels) received over a fixed

bandwidth to a central processing location where they are combined to detect and track targets.

In this example the target state space S is the interval $[0\text{m}, 100\text{m}]$ and the prior is uniform over this interval. There are two sensors. Sensor 1 is located at $x = 0\text{m}$ and sensor 2 at $x = 100\text{m}$. For this example, we suppose that we know the target of interest is a truck and that its radiated sound level is 90 dB (at 10m). (If we do not know the sound level, this will have to be estimated, too.)

If the target is located at x, let $s(l, x)$ be the signal level in decibels received at a sensor at position l. This level will be attenuated due to the spherical spreading of the sound. (We ignore other attenuation factors such as absorption.) Specifically,

$$s(l, x) = 90 - 20 \ln\left(|x - l|/10\right) \text{ for } l = 0,100 \text{ and } 0 < x < 100$$

Figure 2.15 shows a plot of the received signal level at sensors 1 and 2 as a function of target position x.

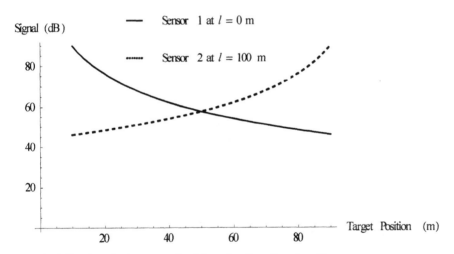

Figure 2.15 Signal received at sensors 1 and 2 as a function of target position.

In addition to the signal received at a sensor, there is random background noise that has a Gaussian distribution with mean 0 and standard deviation $\sigma = 4\,\text{dB}$. Suppose that Y_i is a measurement from sensor i. Then

$$Y_i = s\left(l(i), x\right) + \varepsilon_i$$

where $l(i)$ is the location of sensor i, $\varepsilon_i \sim \mathcal{N}(0,\sigma^2)$ and ε_i is independent of the noise at the other sensor. If the measurement is $Y_i = y_i$, then the likelihood function is

$$L_i(y_i \mid x) = \left(2\pi\sigma^2\right)^{-1/2} \exp\left(-\frac{1}{2\sigma^2}(y_i - s(l(i),x))^2\right) \qquad (2.7)$$

Suppose sensors 1 and 2 obtain measurements $y_1 = 65$ dB and $y_2 = 60$ dB. Figures 2.16 and 2.17 show the likelihood functions for sensor 1 and sensor 2. Figure 2.18 shows the combined likelihood for both sensors, which is the pointwise product of the two likelihood functions. In Figure 2.18, we see that the result of combining the likelihood functions is to produce a peak near 40 m. Since the prior is uniform, the posterior is proportional to the combined likelihood function in Figure 2.18.

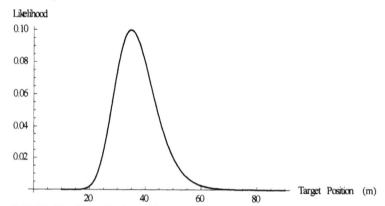

Figure 2.16 Likelihood function for the first measurement from sensor 1 located at $x = 0$.

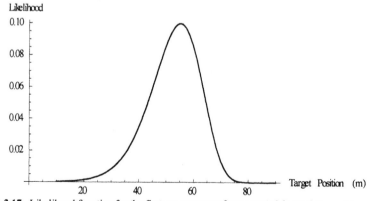

Figure 2.17 Likelihood function for the first measurement from sensor 2 located at $x = 10$.

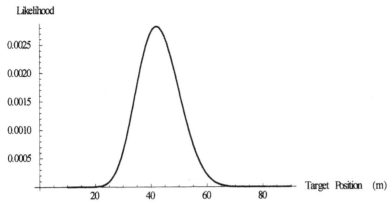

Figure 2.18 Combined likelihood function for the first measurements from sensors 1 and 2.

Suppose that the sensors receive a second set of measurements: 63 dB at sensor 1 and 59 dB at sensor 2 with the background noise being independent of that for the first measurements. Figure 2.19 shows the result of both sets of measurements. The peak is now a bit sharper.

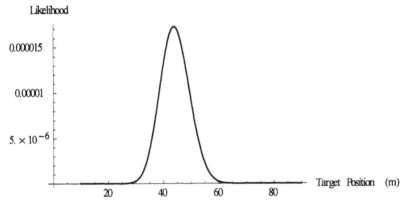

Figure 2.19 Combined likelihood function for both measurements from sensors 1 and 2.

2.3.8 Summary

This section demonstrates how likelihood functions can be used to combine different kinds of measurements such as bearings and position estimates and measurements from different types of sensors such as radar and IR. We have shown how detection and nondetection information can be included in target state estimates through the use of likelihood functions and combined with measurements such as bearings. Finally, we have shown how likelihood functions can be used to estimate target location from basic sensor measurements such as signal plus noise. The examples presented in this section demonstrate the flexibility and power of using likelihood functions to incorporate information from disparate types of sensors and sources into estimates of target state.

References

[1]   McGrayne, S. B., *The Theory That Would not Die*, New Haven, CT: Yale Press, 2011.

[2]   Shafer, G. A., *A Mathematical Theory of Evidence*, Princeton, NJ: Princeton University Press, 1976.

[3]   Zadeh, L. A., *Fuzzy Sets and Systems*, Amsterdam: North Holland, 1978.

[4]   Goodman, I. R. R., R. P. S. Mahler, and H. T. Nguyen, *Mathematics of Data Fusion*, Boston: Kluwer Academic Publishers, 1997.

[5]   Pearl, J. A., *Probabilistic Reasoning in Intelligent Systems,* San Mateo, CA: Morgan Kaufman, 1988.

[6]   Hogg, R. V., A. T. Craig, and J. W. McKean, *Introduction to Mathematical Statistics* (7th *Ed.*), Upper Saddle River, NJ: Pearson Education, 2005.

[7]   Lehmann, E. L., and J. P. Romano, *Testing Statistical Hypotheses* (3rd Ed.), New York: Springer 2005.

[8]   Lehmann, E. L., *Theory of Point Estimation,* New York: John Wiley & Sons, 1983.

[9]   Stuart, A., *Kendall's Advanced Theory of Statistics* (3 volumes), New York, Wiley, 2010.

[10] Berger, J. O., *Statistical Decision Theory and Bayesian Analysis* (2nd Ed.), New York: Springer-Verlag, 1985.

[11] DeGroot, M. H., *Optimal Statistical Decisions, WCL Edition,* New York: Wiley, 2004.

[12] Box, G. E. P., and G. C. Tiao, *Bayesian Inference in Statistical Analysis,* New York: Wiley-Interscience 1992.

[13] Kadane, J. B., *Principles of Uncertainty,* Boca Raton, FL: CRC Press, 2011.

[14] Bayes, T., "An Essay towards solving a problem in the doctrine of chances," *Phil. Trans. Roy. Soc.,* Vol. 53, 1783, pp. 370-418, 1783.

[15] Press, S. J., *Bayesian Statistics: Principles, Models, and Applications,* New York: Wiley, 1989.

Chapter 3

Single Target Tracking

In this chapter we consider the single target tracking problem as a Bayesian inference problem. We assume that there is one target known to be present so that the tracking problem becomes one of estimating the state of that target. In Chapters 4–6, we consider multiple target problems where the number of targets present may be unknown and changing over time. In Chapter 7, we consider the case where there is at most one target present and specifically address the question of when to call a target present.

Section 3.1 provides the basic Bayesian filtering equations for the recursive computation of the posterior distribution on target state when that state is changing over time. We show that this recursive approach can be applied whenever the target's motion model is Markovian and the observational errors are independent conditioned on the target's path. In Section 3.2, we show that Kalman filtering is a special case of Bayesian filtering that applies when certain linear-Gaussian assumptions hold. In Section 3.3, we discuss how to implement the recursive Bayesian filtering equations with a particle filter.

Recursive Bayesian filters are flexible; the assumptions required for the validity of this recursive approach are not restrictive. These filters can process widely disparate types of data, basically any data for which a likelihood function can be written. Such data include typical contact data in the form of elliptical containment regions or lines-of-bearing as well as "negative information" about the failure of a sensor to detect the target. The likelihood function can account for detailed knowledge of the sensor performance as, for example, with conical bearings that are characteristic of linear arrays of sensors.

So far we have been discussing information and data that are revealed during the tracking process. We will call this type of data "dynamic data." Not all data that may affect conclusions about a target's state will be of this form. For example, suppose we are tracking a ship. We know that the ship will not travel onto land. This is information in a very real sense. If this information is not represented to the filter, it can project the target's solution onto the land. This often happens when using Kalman filters in confined areas. Thus, the existence and whereabouts of land is information, and the manner in which we use this information will affect our conclusions.

The location of land (or more generally shallow water) is information of a different sort than the dynamic information revealed by the sensors. We know where the land is throughout the data collection period. It is static rather than dynamic information. Other types of static information include data on historical patterns of operation and knowledge of the target's intentions, such as the fact that it avoids certain regions of the ocean and prefers others. In Section 3.1.6, we show how to modify the motion model to account for static information.

3.1 BAYESIAN FILTERING

This section develops a recursive form of Bayes' theorem called recursive Bayesian filtering. This procedure is often mentioned in standard texts and papers on tracking and data fusion; see for example [1–7]. However, until recently, the numerical difficulties in implementing the recursion have resulted in emphasis being placed on extended Kalman filtering, weighted Gaussian sum approaches, and nonlinear extensions of Kalman filters involving interactive multiple motion models [2]. The availability of powerful, low-cost computers plus the popularization of particle filtering [8] have made the implementation of the general Bayesian recursion practical for a wide range of applications.

3.1.1 Recursive Bayesian Filtering

Bayesian filtering is the application of Bayesian inference to the problem of tracking a single target. It is based on the mathematical theory of probabilistic filtering described by Jazwinski [9]. We consider the situation where the target motion is modeled in continuous time, but the observations are received at discrete (possibly random) times. This is called continuous-discrete filtering by Jazwinski.

3.1.1.1 Problem Definition

The tracking problem considered in this chapter takes place in a region $\mathcal{R}$ that defines the kinematic region of interest for this problem. The region $\mathcal{R}$ may have boundaries defined in terms of bounds on spatial coordinates or on other kinematic characteristics of the target such as speed. We assume that there is one target present in $\mathcal{R}$ so that the tracking problem becomes one of estimating the state of that target. We also assume there are no false alarms. These will be considered in Chapter 4.

Target State Space

Let S be the state space of the target. Typically, the target state will be a vector of components where some of the components are kinematic and may include position, velocity, and possibly acceleration. There can be additional components that may be related to the identity or other features of the target. For example, if one of the components specifies target type, then that component may also specify information such as radar cross section, coefficients of drag, and mass for airborne targets; in the case of naval targets, it might specify radiated noise levels at various frequencies and motion characteristics (e.g., maximum speeds). The state space will vary from problem to problem depending on the nature of the target to be tracked and the sensor (or other) information to be received. In order to use the recursion given in this section, there are additional requirements on the target state space. The state space must be rich enough that (1) the target's motion is Markovian in the chosen state space and that (2) the sensor likelihood functions depend only on the state of the target at the time of the observation.

Prior Information

Let $X(t)$ be the (unknown) target state at time t. We start the problem at time 0 and are interested in estimating $X(t)$ for $t \geq 0$. The prior information about the target is represented by a stochastic process $\{X(t); t \geq 0\}$. Sample paths of this process correspond to possible target paths through the state space, S. The state space S has a probability distribution associated with it. If S is discrete, this distribution is discrete. If S is continuous (e.g., if S is the plane), then this distribution is represented by a density. Integration with respect to this distribution will be indicated by ds. If the distribution is discrete, then integration becomes summation.

Sensors

There is a set of sensors that report observations at an ordered, discrete sequence of (possibly random) times. These sensors may be of different types and report different information. The set can include radar, sonar, optical, or other types of sensors. The sensors may report only when they have a contact or on a regular basis. Each sensor may have a different measurement space. So, observations from sensor j take values in the measurement space M_j. We assume that we know the probability distribution on M_j of the response of sensor j conditioned on the value of the target state s. This relationship is captured in the likelihood function for that sensor. The relationship between the sensor response and the target state s may be linear or nonlinear, and the probability distribution may be

Gaussian or non-Gaussian. This is in contrast to Kalman filtering, which requires these relationships to be linear and probability distributions to be Gaussian.

Likelihood Functions

Suppose that by time t we have obtained observations at the set of times $0 \le t_1 < \ldots < t_K \le t$. To allow for the possibility that we may receive more than one sensor observation at a given time, we let Y_k be the set of sensor observations received at time t_k. Let y_k denote a value of the random variable Y_k. We assume that we can compute the likelihood function

$$L_k(y_k \mid s) = \Pr\{Y_k = y_k \mid X(t_k) = s\} \text{ for } s \in S \qquad (3.1)$$

The computation in (3.1) can account for correlation among sensor responses if that is required.

Let

$$\mathbf{Y}(t_K) = (Y_1, Y_2, \ldots, Y_K) \text{ and } \mathbf{y} = (y_1, \ldots, y_K)$$

Define

$$L(\mathbf{y} \mid s_1, \ldots, s_K) = \Pr\{\mathbf{Y}(t_K) = \mathbf{y} \mid X(t_1) = s_1, \ldots, X(t_K) = s_K\}$$

We assume that the likelihood of the data $\mathbf{Y}(t_K)$ received through time t_K depends only on the target states at the times $\{t_1, \ldots, t_K\}$ so that

$$\Pr\{\mathbf{Y}(t_K) = \mathbf{y} \mid X(u) = s(u), 0 \le u \le t\} = L(\mathbf{y} \mid s(t_1), \ldots, s(t_K)) \qquad (3.2)$$

Posterior

Define

$$q(s_1, \ldots, s_K) = \Pr\{X(t_1) = s_1, \ldots, X(t_K) = s_K\}$$

to be the prior probability (density) that the process $\{X(t); t \ge 0\}$ passes through the states $s_1, \ldots, s_K$ at times $t_1, \ldots, t_K$. Let the function $p(t_K, \cdot)$ be the posterior probability (density) function for $X(t_K)$ given $\mathbf{Y}(t_K) = \mathbf{y}$. Then

$$p(t_K, s_K) = \Pr\{X(t_K) = s_K \mid \mathbf{Y}(t_K) = \mathbf{y}\} \qquad (3.3)$$

Note that we have suppressed the dependence of p on $\mathbf{y}$. In mathematical terms, our problem is to compute this posterior probability function. Recall that from the point of view of Bayesian inference, the posterior distribution on target state represents our knowledge of the target state. All estimates of target state derive from this posterior.

3.1.1.2 Computing the Posterior

We compute the posterior by the use of Bayes' theorem as follows:

$$
\begin{aligned}
p(t_K, s_K) &= \frac{\Pr\{\mathbf{Y}(t_K) = \mathbf{y} \text{ and } X(t_K) = s_K\}}{\Pr\{\mathbf{Y}(t_K) = \mathbf{y}\}} \\
&= \frac{\int L(\mathbf{y} \mid s_1, \ldots, s_K) q(s_1, s_2, \ldots, s_K) ds_1 ds_2 \cdots ds_{K-1}}{\int L(\mathbf{y} \mid s_1, \ldots, s_K) q(s_1, s_2, \ldots, s_K) ds_1 ds_2 \cdots ds_K}
\end{aligned}
\tag{3.4}
$$

Note that we do not integrate over s_K in the numerator of (3.4). Computing $p(t_K, s_K)$ can be quite difficult. The method of computation depends upon the functional forms of q and L. The two most common ways are batch computation and the recursive method.

Batch Method

In the batch method, one computes the integrals in (3.4) at time t_K when all the observations have been obtained. This is in comparison to recursive procedures where one computes the posterior target state distribution after each observation. The posterior is then combined with the next observation to compute a new posterior. This type of recursive computation requires some special assumptions that are discussed below.

Most often a Monte Carlo evaluation of the integrals in (3.4) is performed by replacing the stochastic process $\{X(t); 0 \leq t \leq T\}$ by a finite number J of sample paths $\{x^j(t); 0 \leq t \leq T\}$ for $1 \leq j \leq J$ drawn from the prior process. Usually the paths are chosen so that each has probability $1/J$. One then computes

$$
w^j(t_K) = p\left(t_K, x^j(t_K)\right) = \frac{L\left(\mathbf{y} \mid x^j(t_1), \ldots, x^j(t_K)\right)}{\sum_{j'=1}^{J} L\left(\mathbf{y} \mid x^{j'}(t_1), \ldots, x^{j'}(t_K)\right)} \quad \text{for } 1 \leq j \leq J \tag{3.5}
$$

as a discrete approximation to the posterior distribution at time t_K. The discrete approximation consists of a set of pairs $\{(x^j(t_K), w^j(t_K)); j = 1, \ldots, J\}$ of states $x^j(t_K)$ and probability masses $w^j(t_K)$ located at the states for $1 \leq j \leq J$. An advantage of the batch method is that Markov assumptions are not needed as they are for the recursive method below. Another is that one obtains a distribution on the track history (path) of the target.

If the likelihood function L in (3.5) factors so that for $1 \le j \le J$

$$L\left(\mathbf{y} = (y_1, \dots, y_K) \mid x^j(t_1), \dots, x^j(t_K)\right) = \prod_{k=1}^{K} L_k\left(y_k \mid x^j(t_k)\right) \tag{3.6}$$

then the particle weights $w^j(t_K)$ in (3.5) can be updated recursively in the following sense. Suppose that one has generated J sample paths to cover the complete time interval $[0,T]$ of interest and has computed the posterior path weights in (3.5) for $\mathbf{y} = (y_1, \dots, y_K)$. If an additional observation $Y_{k'} = y_{k'}$ is obtained for time $t_{k'}$, then one can compute

$$L_{k'}\left(y_{k'} \mid x^j(t_{k'})\right) \text{ for } 1 \le j \le J$$

and

$$\tilde{w}^j = \frac{L_{k'}\left(y_{k'} \mid x^j(t_{k'})\right) w^j(t_K)}{\sum_{j'=1}^{J} L_{k'}\left(y_{k'} \mid x^{j'}(t_{k'})\right) w^{j'}(t_K)}$$

to obtain the new posterior probabilities on the paths. Note that $t_{k'}$ can be either earlier or later than t_K. However, $t_{k'}$ must still be in the interval $[0,T]$. This method allows for time-late data (i.e., observations that arrive out of order, in particular after the time of the last observation that has been processed).

Recursive Method

If the likelihood function factors as in (3.6) and the stochastic process $\{X(t); t \ge 0\}$ is Markovian on the state space S, then $p(t_K, s_K)$ may be computed recursively.

We rewrite (3.6) as

$$L(\mathbf{y} \mid s_1, \dots, s_K) = \prod_{k=1}^{K} L_k(y_k \mid s_k) \tag{3.7}$$

The assumption in (3.7) means that the measurements (or observations) at time t_k depend only on the target state at the time t_k. This is not automatically true. For example, if the target state space is position only and the observation is a velocity measurement, this observation will depend on the target state over some time interval near t_k. The remedy in this case is to add velocity to the target state space. There are other observations, such as failure of a sonar sensor to detect an underwater target over a period of time, for which the remedy is not so easy or obvious. Failure to detect may depend on the whole past history of target positions, and perhaps velocities too.

Define the Markov transition function

$$q_k(s_k \mid s_{k-1}) = \Pr\{X(t_k) = s_k \mid X(t_{k-1}) = s_{k-1}\} \quad \text{for } k \geq 1 \tag{3.8}$$

and let $q_0(\cdot)$ be the probability (density) function for $X(0)$. By the Markov assumption

$$q(s_1,\ldots,s_K) = \int_S \prod_{k=1}^{K} q_k(s_k \mid s_{k-1}) q_0(s_0) ds_0 \tag{3.9}$$

Substituting (3.7) and (3.9) into (3.4), we see that $p(t_K, s_K)$ is proportional to

$$\int L_K(y_K \mid s_K) \left[\prod_{k=1}^{K-1} L_k(y_k \mid s_k)\right] q_K(s_K \mid s_{K-1}) \left[\prod_{k=1}^{K-1} q_k(s_k \mid s_{k-1})\right] q_0(s_0) ds_0 \cdots ds_{K-1}$$

$$= L_K(y_K \mid s_K) \int q_K(s_K \mid s_{K-1}) \left[\int \prod_{k=1}^{K-1} L_k(y_k \mid s_k) q_k(s_k \mid s_{k-1}) q_0(s_0) ds_0 \cdots ds_{K-2}\right] ds_{K-1}$$

so that

$$p(t_K, s_K) = \frac{1}{C} L_K(y_K \mid s_K) \int_S q_K(s_K \mid s_{K-1}) p(t_{K-1}, s_{K-1}) ds_{K-1} \tag{3.10}$$

where C is the constant (independent of s_K) that normalizes $p(t_K, \cdot)$ to a probability distribution and is equal to the denominator of (3.4).

3.1.1.3 Single Target Recursion

Equation (3.10) provides a recursive method of computing $p(t_k, \cdot)$. Specifically, we obtain the basic recursion for single target tracking.

Bayesian Recursion for Single Target Tracking

Initial distribution	$p(t_0, s_0) = q_0(s_0)$ for $s_0 \in S$	(3.11)

For $k \geq 1$ and $s_k \in S$,

Motion update	$p^-(t_k, s_k) = \int q_k(s_k \mid s_{k-1}) p(t_{k-1}, s_{k-1}) ds_{k-1}$	(3.12)
Measurement likelihood	$L(y_k \mid s_k) = \Pr\{Y_k = y_k \mid X(t_k) = s_k\}$	(3.13)
Information update	$p(t_k, s_k) = \dfrac{1}{C} L_k(y_k \mid s_k) p^-(t_k, s_k)$	(3.14)

The notation $p^-(t_k, \cdot)$ indicates the target state distribution given the observations $(y_1, \ldots, y_{k-1})$ and motion updated to time t_k, but prior to the incorporation of the observation at time t_k. The motion update in (3.12) accounts for the transition of target state from time t_{k-1} to t_k. Transitions can represent not only the physical motion of the target but also changes in other state variables. If there has been no observation at time t_k, then there is no information update, only a motion update.

The information update in (3.14) is accomplished by pointwise multiplication of $p^-(t_k, s_k)$ by the likelihood function $L_k(y_k \mid s_k)$. Likelihood functions replace and generalize the notion of contacts in a Bayesian tracker. Likelihood functions can represent sensor information such as detections, no detections, position and velocity measurements with Gaussian errors, bearing observations, measured signal-to-noise ratios, and observed frequencies of a signal. Likelihood functions can also represent and incorporate information in situations where the notion of a measurement is not meaningful, (e.g., subjective information can be incorporated by using likelihood functions). Examples of likelihood functions are given in Section 2.3.

Equations (3.13) and (3.14) do not require the observations to be linear functions of the target state, nor do they require the measurement errors or the probability distributions on target state to be Gaussian. We call this *nonlinear, non-Gaussian filtering*.

Except in special circumstances, the Bayesian recursion must be computed numerically. However, the computers available today allow us to compute and display tracking solutions for complex nonlinear trackers. There are two standard approaches to numerical computation of the posterior distribution on target state. In the particle filter approach, the Markov process $\{X(t); t \geq 0\}$ is approximated by simulating (in a recursive fashion) a number J of target paths through the state space S. Typically, each path is assigned prior probability $1/J$. At each time step this path is moved forward to its position s_k at time t_k, and its probability is multiplied by $L(y_k|s_k)$. Moving the paths forward to t_k produces a discrete numerical approximation to (3.12). Multiplication by $L(y_k|s_k)$ and division by C accomplishes (3.14).

In the grid-based approach, one typically discretizes the state space so that (3.12) is computed through the use of discrete transition probabilities. The likelihood functions are also computed on the gridded state space. In order to cover a wide range of possible situations, this approach may use a dynamic grid for the state space where the size and location of the cells in the grid adjust as the distribution on target state moves, expands, or contracts.

We show in Section 3.2 that a Kalman filter tracker represents a special case of the recursion in (3.11) - (3.14) that holds under special assumptions. In particular, the target motion process $\{X(t); t \geq 0\}$ must be Gaussian; it must satisfy a linear stochastic differential (or difference) equation, and the distribution of $X(0)$ must be Gaussian. The observations $Y(t_k)$ must be linear functions of target state with Gaussian errors; that is,

$$Y(t_k) = H_k\, s_k + \varepsilon_k \text{ for } k = 1, \dots, K$$

where H_k is a matrix and ε_k are mean 0, Gaussian random variables such that ε_k is independent of $X(t_k)$ and ε_j for $j \neq k$. Under these assumptions, we show in Section 3.2 that the recursion in (3.11)–(3.14) reduces to the standard Kalman filter. This is called a *linear-Gaussian filter*, and these assumptions are called the *linear-Gaussian* assumptions.

The linear-Gaussian restrictions can be severe in environments where one is required to fuse diverse types of information. These often involve non-Gaussian distributions and nonlinear relationships between the measurements and the target state. When the target state space is in the standard Cartesian coordinates, a line-of-bearing measurement has a nonlinear relation to the target state. Extended Kalman filters handle this problem by approximating nonlinear measurement relations with linear ones. The process requires the choice of a point in the state space about which to perform the linear approximation. The success of this approximation varies depending on how well chosen the point is and how well the measurement relation can be approximated by a linear one. Gaussian motion models are also restrictive, often producing unrealistic motion models.

3.1.2 Prediction and Smoothing

Section 3.1.1 describes the process of Bayesian filtering. In filtering terminology (see [9]), filtering is the process of estimating the present target state. If we are interested in estimating the target's state at some time in the future, this is called prediction.   Estimation of the target's state for a time in the past is called smoothing. This section discusses the extension of the filtering methods presented above to the problems of prediction and smoothing. At the end of this section, we show how smoothing can be combined with likelihood functions to produce a motion model in which targets avoid land.

3.1.3 Recursive Prediction

The solution to the prediction problem is a simple extension of the filtering solution obtained in Section 3.1.1. One applies the motion update step given in (3.12) to the function $p(t,\cdot)$, which gives the posterior distribution on target state at the present time t, to compute the predicted distribution on target state at any time $u > t$.

3.1.4 Recursive Smoothing

The filtering solution given in Section 3.1.1 uses information in the past or present to estimate the target's present state. There are occasions when one has obtained a sequence of observations over the time interval $[0,T]$ on a target, and one wishes to use this information to construct the best estimate of the target's state as a function of time over this interval.   This process is called (fixed interval) smoothing or track reconstruction. To obtain the best estimate of the target's state at time t where $0 < t < T$, one must use the observations obtained after time t as well as before. This becomes a problem of smoothing.

For this discussion we will limit ourselves to situations where the computation can be performed in a recursive manner.

Let $[0, T]$ be the time interval of interest, and suppose that by time T, we have obtained observations $\mathbf{Y}(T) = (y_1, y_2, \ldots, y_K)$ at the set of discrete times $0 \le t_1 \le \ldots \le t_K \le T$. We suppose that the assumptions in Section 3.1.1 hold, in particular that the likelihood functions are independent given the target's track and that the target motion process is Markovian (i.e., that (3.7) and (3.9) hold).

For $0 \le t \le T$ define

$$\bar{p}(t,s \,|\, (y_1, \ldots, y_K)) \equiv \mathbf{Pr}\{X(t) = s \,|\, \mathbf{Y}(T) = (y_1, \ldots, y_K)\}$$
$$= \frac{\mathbf{Pr}\{X(t) = s \text{ and } \mathbf{Y}(T) = (y_1, \ldots, y_K)\}}{\mathbf{Pr}\{\mathbf{Y}(T) = (y_1, \ldots, y_n)\}} \qquad (3.15)$$

Then $\bar{p}(t,s)$ is the probability (density) that $X(t) = s$ given the observations $\mathbf{Y}(T) = (y_1, \ldots, y_K)$.   The probability distribution $\bar{p}(t,\cdot)$ represents our knowledge of the target state at time t given all the observations, including those before, during, and after t. This is the smoothed estimate of $X(t)$. Note that

$\overline{p}(t_K, s | (y_1, \ldots, y_K)) = p(t_K, s)$. Suppose $t = t_k$. Then applying (3.7) and (3.9) to (3.15), we obtain, for some constant C,

$$\overline{p}(t_k, s_k | (y_1, \ldots, y_K))$$
$$= \frac{1}{C} \int \prod_{i=1}^{K} [L_i(y_i | s_i) q_i(s_i | s_{i-1})] q_0(s_0) ds_0 \cdots ds_{k-1} ds_{k+1} \cdots ds_K \quad (3.16)$$

Notice that we do not integrate over s_k on the right-hand side of (3.16). We can rewrite (3.16) as

$$\overline{p}(t_k, s_k | (y_1, \ldots, y_K))$$
$$= \frac{1}{C} L_k(y_k | s_k) \int q_k(s_k | s_{k-1}) \prod_{i=1}^{k-1} [L_i(y_i | s_i) q_i(s_i | s_{i-1})] q_0(s_0) ds_0 \cdots ds_{k-1}$$
$$\times \int \prod_{i=k+1}^{K} [L_i(y_i | s_i) q_i(s_i | s_{i-1})] ds_{k+1} \cdots ds_K$$
$$= \frac{1}{C'} L_k(y_k | s_k) p^-(t_k, s_k) \int \prod_{i=k+1}^{K} [L_i(y_i | s_i) q_i(s_i | s_{i-1})] ds_{k+1} \cdots ds_K$$

where $p^-(t_k, s_k)$ is defined in (3.12) and C' is the appropriate normalizing factor.

3.1.4.1 Smoothed Estimate at Time t_k

For $k = 0, 1, \ldots, K-1$, let

$$f_k(s_k | (y_{k+1}, \ldots, y_K)) = \int \prod_{i=k+1}^{K} [L_i(y_i | s_i) q_i(s_i | s_{i-1})] ds_{k+1} \cdots ds_K$$

Then

$$\overline{p}(t_k, s_k | (y_1, \ldots, y_K)) = \frac{1}{C} L_k(y_k | s_k) p^-(t_k, s_k) f_k(s_k | (y_{k+1}, \ldots, y_K)) \quad (3.17)$$

Observe that $\overline{p}(t_k, s_k | (y_1, \ldots, y_K))$ is proportional to a product of factors corresponding to the effect of past information, $p^-(t_k, s_k)$; future information, $f_k(s_k | (y_{k+1}, \ldots, y_K))$; and present information, $L(y_k | s_k)$. Specifically, $p^-(t_k, s_k)$ is the probability (density) that $X(t_k) = s_k$ conditioned on the past information $Y(t_{k-1}) = (y_1, \ldots, y_{k-1})$. Note that $p^-(t_k, s_k)$ includes the contact information for times $t < t_k$ and is "motion updated" from t_{k-1} to t_k. The factor

$L(y_k \mid s_k)$ is the likelihood function representing the information in the observation $Y_k = y_k$ at the present time, and $f_k(s_k \mid (y_{k+1},...,y_K))$ is proportional to the probability (density) that $X(t_k) = s_k$ given the future observations $Y_{k+1} = y_{k+1},...,Y_K = y_K$.

We have already shown how to compute $p^-(t_k, s_k)$ recursively. We now find a backward recursion for computing $f_k(s_k \mid (y_{k+1},...,y_K))$. In particular, for $s \in S$ and $0 \le k \le K - 1$,

$$f_K(s \mid \varnothing) = 1$$
$$f_k\big(s \mid (y_{k+1},...,y_K)\big) \tag{3.18}$$
$$= \int f_{k+1}\big(s \mid (y_{k+2},...,y_K)\big) L_{k+1}(y_{k+1} \mid s_{k+1}) q_{k+1}(s_{k+1} \mid s) \, ds_{k+1}$$

Equations (3.17) and (3.18) provide a recursive method for computing $\bar{p}\big(t_k, s_k \mid (y_1,...,y_K)\big)$ for $k = 1,...,K$. This is called a forward-backward smoothing algorithm.

3.1.4.2 Smoothed Estimate at Time t where $t_k < t < t_{k+1}$

Suppose that t is a time such that $t_k < t < t_{k+1}$. To compute $\bar{p}(t,s\mid(y_1,...,y_K))$ we define

$$\hat{q}(t_b,s_b \mid t_a,s_a) \equiv \mathbf{Pr}\{X(t_b) = s_b \mid X(t_a) = s_a\} \text{ for } t_a < t_b \text{ and } s_a, s_b \in S$$

Then by an argument similar to the one leading to (3.17), we have for $t_k < t < t_{k+1}$

$$\bar{p}(t,s \mid (y_1,...,y_K)) =$$
$$\frac{1}{C} \int \hat{q}(t,s \mid t_k, s_k) p^-(t_k, s_k) \, ds_k$$
$$\times \int f_{k+1}\big(s_{k+1} \mid (y_{k+2},...,y_K)\big) L_{k+1}(y_{k+1} \mid s_{k+1}) \hat{q}(t_{k+1}, s_{k+1} \mid t, s) \, ds_{k+1} \tag{3.19}$$

3.1.4.3 Mean of the Smoothed Estimate

The function $\bar{p}(t,s\mid(y_1,...,y_K))$ for $0 \le t \le T$ gives the probability distribution representing the smoothed estimate of the target's state at time t given the observations $\mathbf{Y}(T) = (y_1,...,y_K)$. Let

$$\bar{\mu}(t) = \int_S s\,\bar{p}\big(t,s|(y_1,\ldots,y_K)\big)\,ds \text{ for } 0 \le t \le T$$

Then $\bar{\mu}(t)$ is the mean of the smoothed estimate of the target's state at time t. One can plot $\bar{\mu}(t)$ for $0 \le t \le T$ to show an estimated smoothed path for the target given the complete set of observations in $\mathbf{Y}(T)$.

3.1.5 Batch Smoothing

For batch smoothing, one can obtain a discrete approximation to the smoothed posterior for any time $0 \le t \le T$ from the posterior at t_K computed by the batch method described in Section 3.1.1.2. The approximation is the discrete distribution given by $\{(x^n(t), w_n(t_K)); n = 1,\ldots,N\}$ where $w_n(t_K)$ is computed by (3.5) and $x^n(t)$ is the position of nth sample path at time t.

3.1.6 Land Avoidance

One application of the smoothing equations just derived is to the problem of land avoidance. The location of land in a maritime tracking application is information. Many tracking approaches do not account for this piece of information.

In nonlinear smoothing, we can account for the fact that ships avoid land and shallow water by using a set of land avoidance likelihood functions that in effect modify the motion model of the target as it approaches land. The advantage to doing this by likelihood functions is that these functions can be applied to any underlying motion model. (A similar approach may be used for tracking ground targets traveling over varying terrain.) This means that we do not have to tailor the motion models to make sure they avoid land each time we change the underlying model. It will happen automatically through the use of the land avoidance likelihoods.

The formalism is entirely parallel to that which leads to (3.17). For each time t_k, we define a likelihood function $L_a(y_k|\cdot)$ called the *land-avoidance likelihood function*. This function does not depend on time. The value of this function is small for those states close to land and for those velocities that are likely to put the target close to land before a velocity change takes place. With this in mind, (3.17) provides a way of computing the distribution on target location at any time t_k knowing that the target avoided land in the past, is avoiding it in the present, and will avoid it in the future.

If an observation is obtained from a sensor at any time t_k, its likelihood function is calculated and combined by pointwise multiplication with the land avoidance likelihood function to form a joint likelihood function L_k to be applied at time t_k. In doing this, we are assuming conditional independence of the land avoidance likelihood function. With this addition, the computations leading to (3.17) remain the same as before.

3.1.6.1 Modified Motion Model

The use of future information to condition the future and present motion of the target is equivalent to modifying the transition probabilities of the Markov motion model. Land avoidance provides a good illustration of this equivalence. Let the future likelihood function f_k be defined by (3.18) for $k = 0,\ldots,K$ with $L_k(y_k \mid \cdot) = L_a$ for $k = 0,\ldots,K$. Define a modified initial probability measure q_0^* and modified transition probabilities q_k^* for $k = 1,\ldots,K$ as follows:

$$q_0^*(s_0) = \frac{q_0(s_0)f_0(s_0)L_a(s_0)}{\int_S q_0(s')f_0(s')L_a(s')ds'} \quad \text{for } s_0 \in S$$

$$q_k^*(s_k \mid s_{k-1}) = \frac{q_k(s_k \mid s_{k-1})f_k(s_k)L_a(s_k)}{f_{k-1}(s_{k-1})} \quad \text{for } s_{k-1}, s_k \in S \tag{3.20}$$

Then for any path $(s_0,\ldots,s_K)$, we have

$$q_0^*(s_0)\prod_{k=1}^{K} q_k^*(s_k \mid s_{k-1}) = \frac{q_0^*(s_0)}{f_0(s_0)}\prod_{k=1}^{K} q_k(s_k \mid s_{k-1})L_a(s_k)$$

$$= \frac{1}{C}q_0(s_0)L_a(s_0)\prod_{k=1}^{K} q_k(s_k \mid s_{k-1})L_a(s_k) \tag{3.21}$$

where

$$C = \int_S q_0(s_0)f_0(s_0)L_a(s_0)\,ds_0$$

$$= \int_S q_0(s_0)L_a(s_0)\left[\int \prod_{k=1}^{K} q_k(s_k \mid s_{k-1})L_a(s_k)ds_1 \cdots ds_K\right]ds_0 \tag{3.22}$$

From (3.21) and (3.22), we see that the probability measure on the path $(s_0,\ldots,s_K)$ produced by the modified transition probabilities, q^*, is exactly the conditional probability of the path $(s_0,\ldots,s_K)$ given land avoidance at times $t_0,\ldots,t_K$.

In the case of Gaussian motion models, such as those described in Section 3.2.2.1, [10] shows how knowledge of future target locations or destinations modifies the motion model.

3.1.6.2 Example of Land-Avoidance Likelihood Function

The following is an example of a land-avoidance likelihood function. This one is appropriate for motion models like the one used for submarines in Chapter 1 where the time between velocity changes has an exponential distribution.

Suppose that we have two buffer distances from land

$$r_n = \text{near distance}$$
$$r_f = \text{far distance}$$

We form a buffer zone by assuming that a submarine will not operate closer to land than the near distance, and at the far distance a submarine will be able operate freely without worry about the water depth. Between these two distances, the submarine becomes increasingly cautious as it approaches land and more likely to turn away from land.[1]

Let α be the avoidance function. This is a function such that

$$\alpha(r) = \begin{cases} 0 & \text{for } 0 \le r \le r_n \\ 1 & \text{for } r_f \le r < \infty \end{cases}$$

and $\alpha(r)$ increases monotonically from 0 to 1 as r increases from r_n to r_f. The land-avoidance likelihood function L_a depends on the position and the velocity of the target state. Specifically, let

$R(s) = $ the minimum distance from the position component of state s to land

$\tau(s) = $ mean time between velocity changes given the target is in state s

and define

$$L_a(s) = \alpha(R(s))\alpha\big(R(s + \tau(s)v)\big) \quad \text{for } s \in S \qquad (3.23)$$

where v is the velocity component of the state s. In (3.23) we multiply the land-avoidance function, evaluated at the present state, by a factor that is equal to the land-avoidance function evaluated at the state obtained by projecting the present state ahead at its velocity, v, for a time equal to the mean time between velocity changes for that state. If this projected state is close to land, then this state (i.e.,

---

[1] This is a simplified model. In practice we set a shallow depth and a deep depth with the shallow depth being the minimum water depth at which a submarine will operate and the deep depth being a minimum water depth at which the submarine is comfortable operating. We then calculate depth contours for each of these depths to form a buffer zone that depends on the location of the submarine.

the combination of position and velocity) receives a low likelihood value. Thus position-velocity combinations that tend to put the target near land become unlikely.

3.2 KALMAN FILTERING

This section shows that the standard Kalman filter is a special case of nonlinear tracking. There are two basic requirements to insure the validity of the standard Kalman filter equations beyond the ones required for recursive Bayesian filtering. They concern the motion and measurement assumptions given as follows. Let

$\mathcal{N}(\mu, \Sigma)$ denote a Gaussian distribution with mean μ and covariance Σ

$\eta(\cdot, \mu, \Sigma)$ be the probability density function for $\mathcal{N}(\mu, \Sigma)$

If the random variable X is distributed as $\mathcal{N}(\mu, \Sigma)$, then we write $X \sim \mathcal{N}(\mu, \Sigma)$.

Motion Assumptions

For Kalman filtering, the Markov transition functions must have the property that whenever they are applied to a Gaussian distribution on the target state space, they produce another Gaussian distribution.

Measurement Assumptions

All observations (measurements), Y, must be linear functions of the target state with Gaussian observation (measurement) errors. In particular, an observation Y_k obtained at time t_k must satisfy the following measurement equation

$$Y_k = H_k X(t_k) + \varepsilon_k \tag{3.24}$$

where

$\qquad X(t_k)$ is an l-dimensional column vector

$\qquad Y_k$ is an r-dimensional column vector with $r \leq l$

$\qquad H_k$ is an $r \times l$ matrix

$\qquad \varepsilon_k \sim \mathcal{N}(0, R_k)$ where R_k is r-dimensional

The measurement error ε_k is independent of $X(t_k)$ and ε_j for $j \neq k$.

The measurement assumptions guarantee that if the distribution of $X(t_k)$ is Gaussian prior to the observation, then it will be Gaussian posterior to the observation. Thus the motion and measurement assumptions guarantee that a Gaussian prior remains Gaussian under motion and information updates.

3.2.1 Discrete Kalman Filtering

In discrete Kalman filtering the target motion takes place in discrete time increments. Let $(t_1, t_2, \ldots)$ be the sequence of times for which the motion model is defined. The observations are obtained at these times. Often there is a fixed interval of time $\Delta = t_k - t_{k-1}$ for all $k \geq 1$, but this is not necessary.

3.2.1.1 Motion Model

The motion model has the following form

$$X(t_k) = F_k X(t_{k-1}) + \varphi_k + w_k \text{ for } k = 1, 2, \ldots \quad (3.25)$$

where

F_k is an $l \times l$ matrix

φ_k is an l-dimensional vector

$w_k \sim \mathcal{N}(0, Q_k)$

$X(t_0) \sim \mathcal{N}(\mu_0, P_0)$

The vector φ_k is the *drift term* and F_k is called the *transition matrix*. The random increment w_k is independent of $X(t_{k-1})$ and w_j for $j \neq k$. From (3.25) we see that if $X(t_{k-1})$ is Gaussian, then $X(t_k)$ is a linear combination of independent Gaussian random variables and is therefore Gaussian. Since $X(t_0)$ is Gaussian, we have

$$X(t_k) \sim \mathcal{N}(\mu_k, P_k) \text{ for } k = 1, 2, \ldots$$

where the mean and covariance, μ_k and P_k, may be computed recursively as follows:

$$\begin{aligned} \mu_k &= F_k \mu_{k-1} + \varphi_k \\ P_k &= F_k P_{k-1} F_k^T + Q_k \end{aligned} \quad (3.26)$$

From (3.25), we see that the motion model is Markov with transition density functions,

$$q_k(x_k \mid x_{k-1}) = \eta(x_k, F_k x_{k-1} + \varphi_k, Q_k) \text{ for } k \geq 1 \quad (3.27)$$

and

$$q_0(x) = \eta(x, \mu_0, P_0)$$

In the following, we give three examples of discrete time motion models.

Motion Model 1: Deterministic with Unknown Initial State

A particularly simple example occurs when the target is assumed to have a constant velocity and $w_k = 0$ for $k \geq 1$. Let the target state space be position and velocity in the usual Cartesian coordinates. Let $x = (z, v)^T$ where z and v are the position and velocity components of the target state, and let

$$\varphi_k = 0 \text{ and } F_k = \begin{pmatrix} 1 & \Delta_k \\ 0 & 1 \end{pmatrix} \text{ where } \Delta_k = t_k - t_{k-1} \text{ for } k \geq 1 \qquad (3.28)$$

Then

$$\begin{pmatrix} z_k \\ v_k \end{pmatrix} = F_k \begin{pmatrix} z_{k-1} \\ v_{k-1} \end{pmatrix} = \begin{pmatrix} z_{k-1} + \Delta_k v_{k-1} \\ v_{k-1} \end{pmatrix}$$

In this case the target's motion is deterministic once the Gaussian-distributed initial state $(z_0, v_0)^T$ is known.

Motion Model 2: Independent Increments for the Velocity Process

As a second example, suppose that F_k and φ_k are given by (3.28) and that

$$w_k = \begin{pmatrix} 0 \\ w_k' \end{pmatrix}$$

where $\{w_k' : k \geq 1\}$ is a sequence of independent, identically-distributed, zero-mean Gaussian random variables with values in the velocity component of the state space. Then the motion model becomes

$$\begin{pmatrix} z_k \\ v_k \end{pmatrix} = F_k \begin{pmatrix} z_{k-1} \\ v_{k-1} \end{pmatrix} + \begin{pmatrix} 0 \\ w_k' \end{pmatrix} = \begin{pmatrix} z_{k-1} + \Delta_k v_{k-1} \\ v_{k-1} + w_k' \end{pmatrix}$$

In this case the position process is the integral of a velocity process with independent Gaussian increments. This allows the target to change velocity, but the process may produce unreasonably large velocities if no measurements are obtained over a long period of time. If the covariance of w'_k is "small" for all k, then this is often called the *nearly constant velocity model*.

Motion Model 3: Discrete-Time Integrated Ornstein-Uhlenbeck Process

In this model, we specify a "drag coefficient" $\gamma > 0$, and let $\{r_k : k \geq 1\}$ be a sequence of independent, identically distributed, zero-mean Gaussian random variables taking values in the velocity space of models 1 and 2 above. In addition, we let

$$\varphi_k = 0, \ F_k = \begin{pmatrix} 1 & \Delta_k \\ 0 & e^{-\gamma\Delta_k} \end{pmatrix}, \text{ and } w_k = \begin{pmatrix} 0 \\ w'_k \end{pmatrix} \text{ where } w'_k = \left(1 - e^{-\gamma\Delta_k}\right) r_k \text{ for } k \geq 1$$

Then

$$\begin{pmatrix} z_k \\ v_k \end{pmatrix} = F_k \begin{pmatrix} z_{k-1} \\ v_{k-1} \end{pmatrix} + \begin{pmatrix} 0 \\ w'_k \end{pmatrix} = \begin{pmatrix} z_{k-1} + \Delta_k v_{k-1} \\ e^{-\gamma\Delta_k} v_{k-1} + w'_k \end{pmatrix} \text{ for } k \geq 1$$

This is a discrete-time version of the integrated Ornstein-Uhlenbeck (IOU) process discussed in Section 3.2.2.2. In this process, the variance of the velocity distribution is bounded as $t \to \infty$.

Comments on Models 2 and 3

While motion models 2 and 3 above are convenient for use in a Kalman filter, they are not very realistic. In Model 2, the variance of the velocity distribution increases linearly with time, producing unreasonable velocity (and position) distributions unless measurements are received without large time gaps between them. Model 3 solves that problem by using a velocity distribution whose variance is bounded even when no measurements are received. Even with this improvement, the Gaussian distribution on velocities allows for the possibility of unrealistically large velocities. Following the description of the continuous time Gaussian motion models in Section 3.2.2.2, there is a discussion of the appropriateness of those models.

3.2.1.2 Measurement Model

The measurements in the discrete time model are obtained at the times t_k for $k \geq 1$. The measurements follow the model in (3.24). They are linear functions of the target state with Gaussian errors.

3.2.1.3 Discrete Kalman Recursion

We now derive the standard Kalman filter recursion as a special case of the basic tracking recursion in (3.11)–(3.14). Let S be the l-dimensional target state space. As before, we let

$$Y(t_k) = y_k \text{ be the measurement obtained at } t_k \text{ for } k = 1, 2, \ldots$$

$$p^-(t_k, s) = \Pr\{X(t_k) = s \mid \mathbf{Y}(t_{K-1}) = (y_1, y_2, \ldots, y_{K-1})\} \qquad (3.29)$$

$$p(t_k, s) = \Pr\{X(t_k) = s \mid \mathbf{Y}(t_K) = (y_1, y_2, \ldots, y_K)\}$$

Then $p^-(t_k, \cdot)$ is the predicted distribution on target state at time t_k, conditioned on the information received through time t_{k-1}. The function $p(t_k, \cdot)$ is the posterior distribution on target state at time t_k conditioned on the information received through time t_k.

   Suppose that

$$p(t_{k-1}, s) = \eta(s, \mu_{k-1}, P_{k-1}) \text{ for } s \in S$$

Applying the motion update (3.12) to $p(t_{k-1}, \cdot)$ and using the transition function q_k in (3.27) produces the target state distribution updated for motion from time t_{k-1} to t_k, which is equal to $p^-(t_k, \cdot)$ as defined in (3.29). Thus

$$p^-(t_k, s) = \eta(s, \mu_k^-, P_k^-)$$

where

$$\begin{aligned}
\mu_k^- &= F_k \mu_{k-1} + \varphi_k \\
P_k^- &= F_k P_{k-1} F_k^T + Q_k
\end{aligned} \qquad (3.30)$$

   Suppose that $Y_k = y_k$. Since $p(t_k, \cdot)$ is obtained from $p^-(t_k, \cdot)$ by conditioning on $Y_k = y_k$ and Y_k satisfies (3.24), we have from the Gaussian density lemma in the Appendix, that

$$p(t_k, s) = \eta(s, \mu_k, P_k)$$

where

$$\begin{aligned}
\mu_k &= \mu_k^- + P_k^- H_k^T \left(H_k P_k^- H_k^T + R_k \right)^{-1} \left(y_k - H_k \mu_k^- \right) \\
P_k &= P_k^- - P_k^- H_k^T \left(H_k P_k^- H_k^T + R_k \right)^{-1} H_k P_k^-
\end{aligned} \qquad (3.31)$$

We can now state the basic recursion for single target tracking in the form that is used in Kalman filtering. The result is the standard discrete Kalman filter recursion (e.g., see (2.11a)–(2.11e) in [3]). This shows that discrete Kalman filtering is a special case of the Bayesian recursion for single target tracking.

Discrete Kalman Filter Recursion

Initial distribution $\qquad p(0,x) = \eta(x, \mu_0, P_0)$

For $k \geq 1$,

  Motion update

$$\mu_k^- = F_k \mu_{k-1} + \varphi_k$$
$$P_k^- = F_k P_{k-1} F_k^T + Q_k$$
$$p^-(t_k, x) = \eta(x, \mu_k^-, P_k^-)$$

  Measurement likelihood $\qquad L(y_k \mid s_k) = \eta(y_k, H_k s_k, R_k)$

  Information update

$$\mu_k = \mu_k^- + P_k^- H_k^T \left(H_k P_k^- H_k^T + R_k\right)^{-1} \left(y_k - H_k \mu_k^-\right)$$
$$P_k = P_k^- - P_k^- H_k^T \left(H_k P_k^- H_k^T + R_k\right)^{-1} H_k P_k^-$$
$$p(t_k, x) = \eta(x, \mu_k, P_k)$$

The matrix

$$K_G = P_k^- H_k^T (H_k P_k^- H_k^T + R_k)^{-1}$$

is called the Kalman gain. The information update equations can be rewritten as follows in terms of the Kalman gain.

$$\mu_k = \mu_k^- + K_G \left(y_k - H_k \mu_k^-\right)$$
$$P_k = \left(I - K_G H_k\right) P_k^-$$

The difference $y_k - H_k \mu_k^-$ between the actual measurement y_k and the expected measurement $H_k \mu_k^-$ from the motion-updated distribution at time t_k is called the *innovation*. The Kalman gain determines the weight given to the innovation in computing μ_k. From the definition of K_G we can see that the smaller[2] the measurement covariance R_k is compared to $H_k P_k^- H_k^T$, the more weight is given

---

[2] The size of a matrix is measured by its determinant.

to the innovation. In addition, the smaller R_k is compared to $H_k P_k^- H_k^T$, the closer $K_G H_k$ is to I and the smaller P_k becomes. So the smaller R_k is compared to $H_k P_k^- H_k^T$, the larger the weight given to the measurement (i.e., the larger the gain applied to the measurement).

3.2.2 Continuous-Discrete Kalman Filtering

In continuous-discrete Kalman filtering, the target motion takes place in continuous time, but the observations are received at discrete (possibly random) times. This is the standard continuous-discrete formulation used throughout this book. In this section, we describe the class of motion models that are employed for continuous-discrete Kalman filtering and then give a recursive method for motion updating with these models. Subsequently, it is straightforward to develop the continuous-discrete Kalman filter recursion.

3.2.2.1 Class of Motion Models

The motion models are obtained as solutions of linear stochastic differential equations. We assume that the target motion process $\{X(t); t \geq 0\}$ on the l-dimensional state space S satisfies

$$dX(t) = \left[F(t)X(t) + a(t)\right]dt + G(t)\,dw(t) \tag{3.32}$$

where $F(t)$ and $G(t)$ are $l \times l$ deterministic matrices, $a(t)$ is a deterministic l vector, and $w(\cdot)$ is a vector of l mutually independent Wiener processes each with unit variance. This is a natural class of target motion models for a Kalman filter since if $X(0)$ is Gaussian, the solution to (3.32) is a Gaussian Markov process. (See page 111 in [9].) The Gaussian and Markov properties are exactly the ones needed for Kalman filtering.

The functions F, a, and G have natural interpretations. The functions F and a determine the motion and drift of the motion process; G determines the diffusion. To see this, suppose that the target's position at time t is x_t. Then the target's position a short time later, $t + dt$, has a distribution that is Gaussian with mean $\mu(t + dt)$ and covariance $\Sigma(t + dt)$ given by

$$\mu(t + dt) = x_t + \left[F(t)x_t + a(t)\right]dt$$
$$\Sigma(t + dt) = G(t)G(t)^T\,dt$$

Note that the variance of the position distribution grows linearly in time.

Figure 3.1 illustrates the concepts of drift and diffusion. In the case where the target state is the two-dimensional position (x_1, x_2), the distribution of the target's position $x_t + dX(t)$ at time $t + dt$, given it was at x_t at time t, can be represented by an ellipse x centered at

$$\bar{x} = x_t + \left[F(t)x_t + a(t)\right]dt$$

and satisfying

$$\left(x - \bar{x}\right)^T \Sigma^{-1}(t + dt)\left(x - \bar{x}\right) = 4 \tag{3.33}$$

This is the ellipse of points at "distance" 2 units from $\bar{x}$ measured in the normalized distance units determined by the covariance matrix $\Sigma(t + dt)$. This ellipse is called the 2-σ ellipse and is used to graphically represent a bivariate Gaussian distribution as shown in Figure 3.1. The target's position, $x_t + dX(t)$, at time $t + dt$ will lie within this ellipse with probability 0.86. Thus we see that the drift $F(t)x_t + a(t)$ determines the displacement of the mean, and the dispersion $G(t)$ determines the spread of the distribution about the mean displacement.

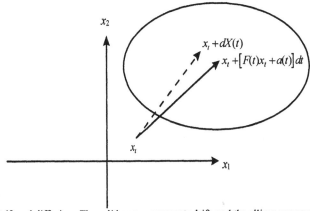

Figure 3.1 Drift and diffusion. The solid arrow represents drift, and the ellipse represents diffusion.

Explicit Solutions

If the drift and diffusion are constant, then we can obtain the following explicit solution to (3.32). Suppose that

$$F(t) = \Gamma, \ a(t) = a, \text{ and } G(t) = G \text{ for } t \geq 0$$

so that (3.32) becomes

$$dX(t) = \left[\Gamma X(t) + a\right]dt + G\,dw(t) \tag{3.34}$$

Under the initial conditions

$$X(0) \sim \mathcal{N}\left(\mu_0, \Sigma_0\right)$$
$$X(0) \text{ is independent of } \left\{w(t); t \geq 0\right\}$$

we can solve (3.34) explicitly using stochastic integration (Corollary (8.2.4) of [11]) to obtain

$$X(t) = e^{\Gamma t}\left[X(0) + \int_0^t e^{-\Gamma u}\left(a\,du + G\,dw(u)\right)\right] \tag{3.35}$$

From (3.35), we can calculate the mean and covariance of $X(t)$ as follows:

$$\mathbf{E}[X(t)] = e^{\Gamma t}\mathbf{E}[X(0)] + \int_0^t e^{\Gamma(t-u)}a\,du = e^{\Gamma t}\mu_0 + \int_0^t e^{\Gamma(t-u)}a\,du \tag{3.36}$$

$$\mathbf{Var}[X(t)] = \mathbf{Var}\left\{e^{\Gamma t}\left[X(0) + \int_0^t e^{-\Gamma u}G\,dw(u)\right]\right\}$$
$$= \mathbf{Var}\left\{e^{\Gamma t}[X(0)]\right\} + \mathbf{Var}\left\{\int_0^t e^{\Gamma(t-u)}G\,dw(u)\right\} \tag{3.37}$$
$$= e^{\Gamma t}\Sigma_0\left(e^{\Gamma t}\right)^T + \int_0^t e^{\Gamma(t-u)}GG^T\left(e^{\Gamma(t-u)}\right)^T du$$

$$\mathbf{Cov}[X(t), X(u)] = \mathbf{Cov}\left\{e^{\Gamma t}X(0), e^{\Gamma u}X(0)\right\}$$
$$+ \mathbf{Cov}\left\{\int_0^t e^{\Gamma(t-r)}G\,dw(r), \int_0^u e^{\Gamma(u-r)}G\,dw(r)\right\}, \tag{3.38}$$
$$= e^{\Gamma t}\Sigma_0\left(e^{\Gamma u}\right)^T + \int_0^{\min(u,t)} e^{\Gamma(t-r)}GG^T\left(e^{\Gamma(u-r)}\right)^T dr$$

Since $X(0)$ and the Wiener process are Gaussian and since the solution in (3.35) is a linear combination of these random variables, it is also Gaussian. Thus, the process $\{X(t); t \geq 0\}$ is Gaussian with the mean and covariance functions as given above. From the representation in (3.35), it also follows that the process $\{X(t); t \geq 0\}$ is Markov.

3.2.2.2 Examples of Continuous-Time Motion Models

We now present three examples of motion models that are obtained from (3.34) and provide the explicit solution for one of them.

Brownian Motion

For this example we take the target state space to be the target's position in the usual two-dimensional Cartesian space. In addition

$$\Gamma = 0, \ a = 0, \ \text{and} \ G = \sigma \mathbf{I}$$

where $\mathbf{I}$ is the 2×2 identity matrix. Then

$$dX(t) = \sigma \, dw(t)$$

and we obtain the standard Wiener process or Brownian motion scaled by σ. For this process the local drift is zero, and the dispersion is constant.

This is not a very satisfactory motion model for most problems because Brownian motion is nowhere differentiable. This means that there is no well-defined velocity for the target. Even worse, the target can move at arbitrarily high average speeds over short time periods. Finally, the "variance" of the position grows linearly over time, which produces unreasonably large dispersions for most problems.

Ornstein-Uhlenbeck Process

As with Brownian motion, we take the state space to be the target's position. Let

$$\Gamma = -\gamma \mathbf{I} \ , \ a = 0, \ \text{and} \ G = \sigma \mathbf{I}$$

where $\gamma > 0$ is called the drag coefficient. Then (3.34) becomes

$$dX(t) = -\gamma X(t)dt + \sigma \, dw(t)$$

This produces the Ornstein-Uhlenbeck process, which can be thought of as Brownian motion with a drag that tends to pull the Brownian particle toward the origin. This drag is proportional to the displacement of the particle from the origin. As a motion model, this process has two problems. First, like Brownian motion it has no velocity. Second, the effect of the drag coefficient γ on the distribution of the process is to produce a limit on the covariance matrix of the process. The covariance of the process approaches this limit asymptotically as time increases. Thus there is a bound on the dispersion of the target's position distribution. In the case where the target state space is two-dimensional, the limit of the mean squared displacement from the origin is σ^2 / γ.

Integrated Ornstein-Uhlenbeck Process

In the IOU process, the state is position and velocity. The position component is obtained by integrating an Ornstein-Uhlenbeck process that models the velocity of the target. Specifically, we take $w(t)$ to be a four-dimensional vector and

$$x = \begin{pmatrix} z \\ v \end{pmatrix} \text{ where } z \text{ is position and } v \text{ is velocity}$$

$$a = 0, \ \Gamma = \begin{bmatrix} 0 & \mathbf{I} \\ 0 & -\gamma\mathbf{I} \end{bmatrix}, \text{ and } G = \begin{bmatrix} 0 & 0 \\ 0 & \sigma\mathbf{I} \end{bmatrix}$$

Then (3.34) becomes

$$d\begin{pmatrix} Z(t) \\ V(t) \end{pmatrix} = \begin{bmatrix} 0 & \mathbf{I} \\ 0 & -\gamma\mathbf{I} \end{bmatrix}\begin{bmatrix} Z(t) \\ V(t) \end{bmatrix} dt + \begin{bmatrix} 0 & 0 \\ 0 & \sigma\mathbf{I} \end{bmatrix} dw(t)$$

This process has a number of virtues as a motion process. It has a well-defined velocity, and we can specify a limit on the mean square speed of the target by specifying σ. In the case where the velocity is two-dimensional, the mean square speed will approach σ^2 / γ in the limit as time increases to infinity. If we start the process with the limiting velocity distribution, then the velocity process $\{V(t); t \geq 0\}$ becomes a stationary Gaussian one with

$$V(t) \sim \mathcal{N}(0, \Sigma_0), \text{ where } \Sigma_0 = \frac{\sigma^2}{2\gamma}\mathbf{I}$$

One can show that the correlation between $V(t)$ and $V(t+s)$ is $e^{-\gamma s}\mathbf{I}$ for $s \geq 0$. This means that we can control the rate at which the target "changes" velocity by the choice of γ.

Using (3.36) and (3.37), we calculate the mean and variance of the IOU process as follows.

$$\mathbf{E}[X(t)] = e^{\Gamma t}\mu_0 \tag{3.39}$$

$$\mathbf{Var}[X(t)] = e^{\Gamma t}\Sigma_0\left(e^{\Gamma t}\right)^T + \sigma^2\begin{pmatrix} b_{11}(t)\mathbf{I} & b_{12}(t)\mathbf{I} \\ b_{21}(t)\mathbf{I} & b_{22}(t)\mathbf{I} \end{pmatrix} \tag{3.40}$$

where

$$e^{\Gamma t} = \begin{pmatrix} \mathbf{I} & \gamma^{-1}\left(1-e^{-\gamma t}\right)\mathbf{I} \\ 0 & e^{-\gamma t}\mathbf{I} \end{pmatrix}$$

$$b_{11}(t) = \gamma^{-2}\left[t - 2\gamma^{-1}\left(1-e^{-\gamma t}\right) + \tfrac{1}{2}\gamma^{-1}\left(1-e^{-2\gamma t}\right)\right]$$

$$b_{12}(t) = b_{21}(t) = \gamma^{-2}\left[\left(1-e^{-\gamma t}\right) - \tfrac{1}{2}\left(1-e^{-2\gamma t}\right)\right]$$

$$b_{22}(t) = \tfrac{1}{2}\gamma^{-1}\left(1-e^{-2\gamma t}\right)$$

Comments on Continuous-Time Gaussian Motion Models

As in the discrete time case, Gaussian motion models are convenient and necessary for Kalman filters. However they are not realistic motion models for most problems for the following reasons.

Brownian motion has two difficulties. First, the sample paths of the process are nowhere differentiable. This means they do not have a well-defined velocity. Even worse the paths have unbounded variation. This means, among other things, that for any time interval $[t_1, t_2]$ there is no constant K such that

$$\frac{\left|X\left(t''\right) - X\left(t'\right)\right|}{\left|t'' - t'\right|} \leq K \text{ for all } t', \, t'' \in [t_1, t_2]$$

which means that over any interval of time there will be arbitrarily steep jumps in position. Second, the variance in the position distribution grows linearly in time if no measurements are received. After a while the spread of the distribution becomes unrealistically large unless a measurement is received.

The Ornstein-Uhlenbeck process bounds the growth in the variance of the position distribution; however, it still suffers from the problem of not having a well-defined velocity process.

The IOU process has a well-defined velocity process (the Ornstein-Uhlenbeck process) whose variance is bounded. However, the sample paths of the velocity process are not realistic. If one wishes to use these models, he must be satisfied with matching certain moments of the process with moments that would correspond to a realistic motion model as shown in Section 1.2, which provides an example of tracking a surface ship. For the IOU process there are two parameters that we can choose, γ and σ. We interpreted $1/\gamma$ as the mean time between velocity changes for a ship, and we noted that the root mean squared speed of the limiting velocity distribution is equal to $\sigma/\sqrt{\gamma}$. If we match this to what we expect the mean speed of a ship to be, this will determine σ. If we are content to match this moment and interpret the parameter $1/\gamma$ as the mean time between velocity changes and are not concerned about the details of the sample path behavior, then the IOU is a reasonable motion model to use.

3.2.2.3 Continuous-Discrete Kalman Recursion

We can now state the continuous-discrete Kalman recursion.

Continuous-Discrete Kalman Filter Recursion

Initial distribution $\qquad p(0,x) = \eta(x, \mu_0, P_0)$

For $k \geq 1$:

Motion update
   Compute the mean μ_k^- and covariance matrix P_k^- of $X^-(t_k)$ from (3.36) and (3.37) by setting

$$\mu_0 = \mu_{k-1}, \Sigma_0 = P_{k-1}, \text{ and } t = t_k - t_{k-1}$$

Measurement likelihood $\qquad L(y_k \mid s_k) = \eta(y_k, H_k s_k, R_k)$

Information update

$$\mu_k = \mu_k^- + P_k^- H_k^T \left(H_k P_k^- H_k^T + R_k \right)^{-1} \left(y_k - H_k \mu_k^- \right)$$

$$P_k = P_k^- - P_k^- H_k^T \left(H_k P_k^- H_k^T + R_k \right)^{-1} H_k P_k^-$$

$$p(t_k, x) = \eta(x, \mu_k, P_k)$$

We have shown above that if we start with a Gaussian distribution, the application of the motion process resulting from the solution of (3.34) is a Gaussian distribution with the mean vector and covariance matrix given by (3.36) and (3.37). These equations provide a recursive method of performing the motion update step of the recursion. The information update step is identical to that in the discrete recursion, and the continuous-discrete Kalman filter recursion now follows.

As we mentioned above, the IOU process is a good Gaussian process to use for a motion model. Using the explicit formulas for motion updating given by (3.39) and (3.40), we can produce a computationally efficient Kalman filter that has a well-defined velocity process that allows for target maneuvers. This filter has been applied by the U.S. Navy to the problems of tracking submarines and surface ships. In these applications, an extended Kalman filter is used to incorporate line-of-bearing information.

3.2.3 Kalman Smoothing

This section presents two recursions for computing smoothed Kalman filter estimates of the targets state at time t_k given the measurements $\mathbf{Y}_K = (Y_1, \ldots, Y_K)$. The first one is obtained directly from the recursive smoothing equation (3.17). The second is the computationally more convenient Rauch-Tung-Streibel (RTS) smoother [12]. See Chapter 7 of [13] or Chapter 5 of [14] for a further discussion of Kalman smoothers.

3.2.3.1 Forward-Backward Smoother

For the convenience of the reader we display (3.17) and (3.18) below

$$\overline{p}\big(t_k, s_k \mid (y_1, \ldots, y_K)\big) = \frac{1}{C} L_k(y_k \mid s_k) p^-(t_k, s_k) f_k\big(s_k \mid (y_{k+1}, \ldots, y_K)\big) \qquad (3.17)$$

where

$$
\begin{aligned}
&f_K(s \mid \varnothing) = 1 \\
&f_k\big(s \mid (y_{k+1}, \ldots, y_K)\big) \\
&\quad = \int f_{k+1}\big(s \mid (y_{k+2}, \ldots, y_K)\big) L_{k+1}(y_{k+1} \mid s_{k+1}) q_{k+1}(s_{k+1} \mid s) \, ds_{k+1}
\end{aligned}
\qquad (3.18)
$$

From (3.17) we can see that under the linear-Gaussian assumptions of a Kalman filter, the first two factors on the right-hand side of the equation are $p^-(t_k, \cdot)$, the motion-updated Gaussian probability density at time t_k given the observations $\mathbf{Y}(t_{k-1}) = (y_1, \ldots, y_{k-1})$, and $L_k(y_k \mid \cdot)$, the likelihood function for the measurement at time t_k. As a result the product of these two factors is proportional to the Gaussian density of the forward Kalman solution at time t_k. Section 3.2.3.3 shows that $f_k(s_k \mid (y_{k+1}, \ldots, y_K))$ is proportional to a Gaussian density whose mean and covariance are the output of a time-reversed Kalman filter operating on the contacts $Y(t_{k+1}) = y_{k+1}, \ldots, Y(t_K) = y_K$. Thus the probability density of the smoothed solution at time t_k is the product of two Gaussian densities. This is equivalent to having a Gaussian prior on the target state and a measurement with a Gaussian error distribution. The resulting posterior is a Gaussian distribution with mean and covariance equal to weighted averages of the means and covariances corresponding to these two densities as given in (3.43). (See Chapter 9 of [15].) This yields the following forward-backward smoothed solution under the assumption that F_k is invertible.

Forward

Compute μ_k and P_k using the standard (forward) Kalman filter recursion for $k = 1, \ldots, K$.

Backward

Compute $\hat{\mu}_k$ and $\hat{P}_k$ for $k = K-1, \ldots, 1$ by applying the Kalman filter recursion to the reversed motion model

$$\hat{X}(t_{k-1}) = F_k^{-1}\left(\hat{X}(t_k) - \varphi_k\right) + \hat{w}_{k-1} \text{ for } k = K, \ldots, 2$$

where $\hat{w}_{k-1} \sim \mathcal{N}(0, F_k^{-1}Q_{k-1}(F_k^{-1})^T)$.

Specifically at time t_K the reverse filter is initialized with $\hat{P}_K^{-1} = 0$ and $\hat{\mu}_K = 0$. Then for $k < K$, compute

$$A_{k+1} = \left(\hat{P}_{k+1}^{-1} + H_{k+1}^T R_{k+1}^{-1} H_{k+1}\right)^{-1} \text{ and } \alpha_{k+1} = A_{k+1}\left(\hat{P}_{k+1}^{-1}\hat{\mu}_{k+1} + H_{k+1}^T R_{k+1}^{-1} y_{k+1}\right) \quad (3.41)$$

$$\hat{\mu}_k = F_{k+1}^{-1}\left(\alpha_{k+1} - \varphi_{k+1}\right) \text{ and } \hat{P}_k = F_{k+1}^{-1} A_{k+1}\left(F_{k+1}^T\right)^{-1} + F_{k+1}^{-1}Q_{k+1}\left(F_{k+1}^T\right)^{-1} \quad (3.42)$$

Smoothed Solution

Let $\bar{\mu}_k$ and $\bar{P}_k$ be the mean and covariance of the smoothed solution for $k = K, \ldots, 1$. Then for $k = K, \ldots, 1$, we compute

$$\begin{aligned}
\bar{P}_k &= \left(P_k^{-1} + \hat{P}_k^{-1}\right)^{-1} \\
\bar{\mu}_k &= \bar{P}_k\left(P_k^{-1}\mu_k + \hat{P}_k^{-1}\hat{\mu}_k\right)
\end{aligned} \qquad (3.43)$$

3.2.3.2 Rauch Tung Streibel (RTS) Smoother

Compute μ_k, P_k^-, P_k using the standard (forward) Kalman filter recursion for $k = 1, \ldots, K$. Set $\bar{P}_K = P_K$, and $\bar{\mu}_K = \mu_K$ and use the following recursion to compute

$$A_k = P_k F_{k+1}^T \left(P_{k+1}^- \right)^{-1}$$

$$\bar{\mu}_k = \mu_k + A_k \left(\bar{\mu}_{k+1} - F_{k+1}\mu_k \right) \tag{3.44}$$

$$\bar{P}_k = P_k - A_k \left(P_{k+1}^- - \bar{P}_{k+1} \right) A_k^T$$

for $k = K-1, \ldots, 1$. This is the RTS fixed-interval smoother recursion. Derivations of this recursion may be found in [12] or Chapter 7 of [13].

3.2.3.3 Proof of Forward-Backward Filter Equations

We prove the forward-backward recursion equations by explicitly computing f_k in (3.18) and then $\bar{p}$ in (3.17). Readers who are not interested in the details of this proof should skip this section.

To compute f_k we first assume that $f_{k+1}(s_{k+1}) = \eta(s_{k+1}, \mu_{k+1}, \Sigma_{k+1})$, i.e., f_{k+1} is the density function of a Gaussian distribution with mean μ_{k+1} and covariance Σ_{k+1}. We will prove this is true at the end of this section. To simplify our notation we do not show the conditioning of f_k on $(y_{k+1}, \ldots, y_K)$. Since the measurement $Y_{k+1} = y_{k+1}$ satisfies (3.24), we have

$$L_{k+1} \left(y_{k+1} \mid s_{k+1} \right) = \eta \left(y_{k+1}, H_{k+1} s_{k+1}, R_{k+1} \right)$$

Since the motion model satisfies (3.25), it follows that

$$q_{k+1} \left(s_{k+1} \mid s_k \right) = \eta \left(s_{k+1}, F_{k+1} s_k + \varphi_{k+1}, Q_{k+1} \right)$$

As a result (3.18) becomes

$$f_k(s_k) =$$
$$\int \eta(s_{k+1}, \mu_{k+1}, \Sigma_{k+1}) \eta \left(y_{k+1}, H_{k+1} s_{k+1}, R_{k+1} \right) \eta \left(s_{k+1}, F_{k+1} s_k + \varphi_{k+1}, Q_{k+1} \right) ds_{k+1} \tag{3.45}$$

The first two factors in the above integral are a Gaussian density for a "prior" distribution and a likelihood function for a measurement that is a linear function of target state with Gaussian error. As a result, the product of these two factors is proportional to the information-updated posterior density in a Kalman filter. Using the information filter version of this update (see Section 7.2 of [2]) we can write

$$\eta\left(s_{k+1},\mu_{k+1},\Sigma_{k+1}\right)\eta\left(y_{k+1},H_{k+1}s_{k+1},R_{k+1}\right)=C\eta\left(s_{k+1},\alpha_{k+1},A_{k+1}\right) \qquad (3.46)$$

for some constant C where

$$A_{k+1}=\left(\Sigma_{k+1}^{-1}+H_{k+1}^{T}R_{k+1}^{-1}H_{k+1}\right)^{-1} \text{ and } \alpha_{k+1}=A_{k+1}\left(\Sigma_{k+1}^{-1}\mu_{k+1}+H_{k+1}^{T}R_{k+1}^{-1}y_{k+1}\right)$$

Using (3.46) we can write (3.45) as

$$f_{k}(s_{k})=\int C\eta\left(s_{k+1},\alpha_{k+1},A_{k+1}\right)\eta\left(s_{k+1},F_{k+1}s_{k}+\varphi_{k+1},Q_{k+1}\right)ds_{k+1} \qquad (3.47)$$

One can show that

$$\eta\left(s_{k+1},F_{k+1}s_{k}+\varphi_{k+1},Q_{k+1}\right)=C'\eta\left(s_{k},F_{k+1}^{-1}\left(s_{k+1}-\varphi_{k+1}\right),F_{k+1}^{-1}Q_{k+1}\left(F_{k+1}^{T}\right)^{-1}\right)$$

for some constant C' by observing that

$$\left(s_{k+1}-F_{k+1}s_{s}-\varphi_{k}\right)^{T}Q_{k+1}^{-1}\left(s_{k+1}-F_{k+1}s_{s}-\varphi_{k}\right)$$
$$=\left[\left(F_{k+1}F_{k+1}^{-1}\right)\left(s_{k+1}-F_{k+1}s_{s}-\varphi_{k}\right)\right]^{T}Q_{k+1}^{-1}\left[\left(F_{k+1}F_{k+1}^{-1}\right)\left(s_{k+1}-F_{k+1}s_{s}-\varphi_{k}\right)\right]$$
$$=\left(F_{k+1}^{-1}\left(s_{k+1}-\varphi_{k}\right)-s_{k}\right)^{T}F_{k+1}^{T}Q_{k+1}^{-1}F_{k+1}\left(F_{k+1}^{-1}\left(s_{k+1}-\varphi_{k}\right)-s_{k}\right)$$
$$=\left(s_{k}-F_{k+1}^{-1}\left(s_{k+1}-\varphi_{k}\right)\right)^{T}\left[F_{k+1}^{-1}Q_{k+1}\left(F_{k+1}^{T}\right)^{-1}\right]^{-1}\left(s_{k}-F_{k+1}^{-1}\left(s_{k+1}-\varphi_{k}\right)\right)$$

We can now write (3.47) as

$$f_{k}(s_{k})=\int CC'\eta\left(s_{k+1},\alpha_{k+1},A_{k+1}\right)\eta\left(s_{k},F_{k+1}^{-1}\left(s_{k+1}-\varphi_{k+1}\right),F_{k+1}^{-1}Q_{k+1}\left(F_{k+1}^{T}\right)^{-1}\right)ds_{k+1}$$

which is the motion-update equation corresponding to the reversed Markov motion model.

$$X_{k}=F_{k+1}^{-1}\left(X_{k+1}-\phi_{k+1}\right)+\tilde{w}_{k+1} \text{ where } \tilde{w}_{k+1}\sim\mathcal{N}\left(0,F_{k+1}^{-1}Q_{k+1}\left(F_{k+1}^{T}\right)^{-1}\right)$$

Thus $f_{k}(s_{k})=\eta(s_{k},\hat{\mu}_{k},\hat{P}_{k})$ where

$$\hat{\mu}_{k}=F_{k+1}^{-1}\left(\alpha_{k+1}-\varphi_{k+1}\right) \text{ and } \hat{P}_{k}=F_{k+1}^{-1}A_{k+1}\left(F_{k+1}^{T}\right)^{-1}+F_{k+1}^{-1}Q_{k+1}\left(F_{k+1}^{T}\right)^{-1} \qquad (3.48)$$

This proves (3.42) and the backward recursion in Section 3.2.3.1 with the proviso that we have assumed that f_k is proportional to a Gaussian density.

To show that this assumption is correct and to complete the proof, we compute the first step of the backward recursion in (3.18). In this step $k = K - 1$ and $f_K = 1$ so that

$$f_{K-1}(s_{K-1}) = \int \eta(y_K, H_K s_{k+1}, R_K) \eta(s_K, F_K s_{K-1} + \varphi_K, Q_K) ds_K$$

The density $f_K = 1$ is equivalent to a Gaussian density with mean $\mu_K = 0$ and precision matrix $\Sigma_K^{-1} = 0$. In this case the information form of the update shows that $\eta(y_K, H_K s_{k+1}, R_K)$ is proportional to a Gaussian density $\eta(s_K, \alpha_K, A_K)$ where

$$A_K = \left(H_K^T R_K^{-1} H_K\right)^{-1} \text{ and } \alpha_K = A_K \left(H_K^T R_K^{-1} y_K\right)$$

It follows that f_{K-1} is proportional to a Gaussian density with mean and covariance given by $\hat{\mu}_{K-1}$ and covariance $\hat{P}_{K-1}$ in (3.48). Thus f_k is proportional to a Gaussian density for $k < K$ and the proof is complete.

3.3 PARTICLE FILTER IMPLEMENTATION OF NONLINEAR FILTERING

Section 3.1 develops the recursion for Bayesian tracking for a single target. Section 3.2 shows how this recursion can be implemented using a Kalman filter under linear-Gaussian assumptions. This section describes an approach called particle filtering that can be used to implement (approximately) the Bayesian tracking recursion when the linear-Gaussian assumptions are not met—in particular when the measurements are not linear functions of target state, the measurement errors are not Gaussian, or the target motion model is not Gaussian.

Particle filters compute the basic recursion by approximating the distribution on target state at time t by a discrete distribution consisting of a finite set of points $x^j(t)$ for $j = 1, \ldots, J$ in the state space with probability $w^j(t)$ attached to the jth point. In the past this approach has been limited by the additional computational load required by a particle filter compared to a Kalman filter. This induced analysts to use nonlinear extensions of Kalman filters such as the extended Kalman filter or interactive multiple model filtering to handle nonlinear tracking problems. By contrast, the powerful and inexpensive computers available today can easily handle the computational load required by a particle filter in most tracking problems. The results in Chapter 6 of [8] show the advantages in terms of tracker performance of using a particle filter in place of a Kalman filter approximation in the case of bearings-only tracking. In bearings-only tracking the measurements (bearings) are nonlinear functions of target state (position and velocity in the usual four-dimensional state space) and the measurement errors are non-Gaussian in the target state space. The example in Section 1.3 shows how to apply a particle filter to this problem.

In outline a particle filter operates sequentially as follows. Suppose that we have computed the particle filter distribution $\{(x^j(t_{k-1}), w^j(t_{k-1})); j = 1, \ldots, J\}$ for time t_{k-1} and have received a measurement $Y_k = y_k$ at time t_k. For each particle $x^j(t_{k-1})$ we draw a value for $x^j(t_k)$ from an importance distribution $\hat{q}(\cdot \mid x^j(t_{k-1}), y_k)$ (see Chapter 3 of [8] or Part I of [16]) to obtain the particle's state at time t_k. We evaluate the likelihood function $L(y_k \mid \cdot)$ at each of these states and compute

$$w^j(t_k) = w^j(t_{k-1}) \frac{L(y_k \mid x^j(t_k)) q(x^j(t_k) \mid x^j(t_{k-1}))}{\hat{q}(x^j(t_k) \mid x^j(t_{k-1}), y_k)}$$

The result is the particle filter distribution $\{(x^j(t_k), w^j(t_k)); j = 1, \ldots, J\}$ at time t_k.

This particle filter is called a Sequential Importance Sampling (SIS) filter. In the special case where the importance distribution is the motion updated particle distribution, that is, when

$$\hat{q}(x^j(t_k) \mid x^j(t_{k-1}), y_k) = q(x^j(t_k) \mid x^j(t_{k-1}))$$

we obtain the sampling importance resampling (SIR) filter.

In the remainder of this section we describe an SIR particle filter implementation of the Bayesian single target tracking recursion. The SIR filter has the virtue that it follows the recursion in (3.11)–(3.14) and works well in many tracking problems. An understanding of the SIR particle filter will allow the reader to design and implement a particle filter for most tracking problems. However, if the need arises, he can implement one of the more advanced particle filters described in [8].

The particles represent sample paths drawn from the target motion process $\{X(t); t \geq 0\}$. When a particle branches (splits), the resulting particles are viewed as having the same path up to the time of the split as the parent particle so that each particle has a complete track history.

3.3.1 Generating Particles

We begin by generating J particles. The jth particle x^j specifies the target state $x^j(t_k)$ for $0 = t_0 < t_1 < t_2 \ldots$ and is generated as follows.

For $j = 1, \ldots, J$:

- Choose $x^j(t_0)$ by making an independent random draw from the distribution q_0;
- For $k > 0$ choose $x^j(t_k)$ by making an independent random draw from the distribution $q_k\left(\cdot \mid x^j(t_{k-1})\right)$;
- Initially particle j is assigned probability $w^j(0) = 1/J$ for $j = 1, \ldots, J$.

Usually, the path of a particle is generated sequentially in time. As the kth measurement is received at time t_k, the states $x^j(t_k)$ are generated for $j = 1, \ldots, J$. The collection of particles generated in this way forms a discrete sample path approximation to the target motion model.

Section 1.3 provides an example of developing a motion model for a tracking problem and generating particles from that model. Since we generate the particle paths by a Monte Carlo procedure, we can use a motion model that produces paths that are reasonable approximations to the possible motion of the target. Section 1.3 discusses this point in some detail.

3.3.2 Particle Filter Recursion

Suppose that at time t_{k-1} we have particles $x^j(t_{k-1})$ with probabilities $w^j(t_{k-1})$ for $j = 1, \ldots, J$ and that we have received a measurement $Y_k = y_k$ at time t_k. We follow the recursion in (3.12)–(3.14).

Motion update. Obtain $x^j(t_k)$ by making an independent random draw from the distribution $q_k\left(\cdot \mid x^j(t_{k-1})\right)$ for $j = 1, \ldots, J$.

Measurement likelihood. Compute the likelihood function

$$L\left(y_k \mid x^j(t_k)\right) = \Pr\left\{Y_k = y_k \mid X(t_k) = x^j(t_k)\right\} \quad \text{for } j = 1, \ldots, J$$

Information update. Compute the posterior probabilities on the particles

$$w^j(t_k) = \frac{w^j(t_{k-1}) L\left(y_k \mid x^j(t_k)\right)}{\sum_{j'=1}^{J} w_{j'}(t_{k-1}) L\left(y_k \mid x^{j'}(t_k)\right)}$$

Resample. Resample the particles as needed to obtain a set of J particles with equal probability. The discussion below describes one way to do this.

3.3.3 Resampling

After a number of sensor measurements have been incorporated using the method described above, we may find that at time t the posterior weights are highly unequal with some paths having many times the weight of others. This typically happens when measurements localize and concentrate the target state distribution. In order to increase the resolution of the tracker and make efficient use of all the particles, we perform a resampling. In the SIR filter, one resamples after each measurement update. The resampling method described below is designed to guarantee that there are exactly J particles after resampling. If this is not important, one can use the Russian roulette and splitting method of resampling described in Section 1.3.

The goal in resampling is to produce J particles with equal probabilities that provide a good representation of the distribution of the target state at time t. With this in mind, we define $M(j) = Jw^j(t)$ for $j = 1,...,J$. We would like to make $M(j)$ copies of particle j, but since $M(j)$ is usually not an integer, we cannot do this exactly. Instead we make approximately $M(j)$ copies of each particle in such a way that we obtain J particles in total. A simple way to do this is to follow the resampling algorithm given below, which is adapted from Table 3.2 of [8].

In this resampling scheme, low-probability particles tend to be discarded and high-probability ones are split into perturbed copies of themselves. The result is J particles each having probability $1/J$ with more particles located in high-probability regions and fewer located in low-probability regions of the state space.

Resampling Algorithm

1. Set $C_0 = 0$ and compute
    $$C_j = \sum_{j'=1}^{j} w^{j'}(t) \text{ for } j = 1,...,J$$
2. Draw u_1 from a uniform distribution over $[0, 1/J]$ and compute
    $$u_m = u_1 + (m-1)/J \text{ for } m = 2,...,J$$
3. For $j = 1,...,J$ do the following
    For m such that $C_{j-1} \leq u_m < C_j$ set $\xi^m = x^j$
4. For $m = 1,...,J$ perturb the state $\xi^m(t)$ at time t of each particle by the method described below to obtain $\hat{\xi}^m$ which differs from ξ^m only in the perturbed state at time t.
5. Set the probability $\hat{w}^j(t)$ of each particle equal to $1/J$ so that $\{(\hat{\xi}^m, \hat{w}^m(t)); m = 1,...,J\}$ becomes the resampled particle distribution.

In step 5 we copy the entire sample path up to time t. Step 6 perturbs the state of the sample path only at time t. Step 6 is often modified so that the first copy of the particle x^j is not perturbed.

3.3.4 Perturbing Target States

In order to increase the resolution and diversity of the particle filter representation of the distribution, we perturb the copies of a particle so that their states are a "little bit" different than the state of the parent particle. If this is not done, the particles can cluster around a small number of states and fail to be representative of the distribution we are trying to estimate. This process is called regularization by a number of authors (e.g., [8, 17]), and the resulting particle filter is called a regularized particle filter.

One way to think about the process of perturbing particles states is that the particles are samples from an underlying smooth probability density. Suppose one estimates this smooth density using kernel functions. There is a substantial literature on this type of density estimation. See for example [18]. In particular suppose that the target state space S is a standard d-dimensional Cartesian space and that we wish to estimate $p(t_k, \cdot)$ by

$$p(t_k, s) \approx \sum_{j=1}^{J} w^j (t_k) K_h \left(s - x^j (t_k) \right) \tag{3.49}$$

where K_h is a kernel function with window size $h > 0$. A standard kernel function is the Gaussian density where

$$K_h(x) = h^{-d} \eta \left(h^{-1} x, 0, \Sigma \right) \text{ for covariance matrix } \Sigma \text{ and } h > 0 \tag{3.50}$$

In this case we are estimating $p(t_k, \cdot)$ in (3.49) by a weighted sum of Gaussian densities centered at the states of the particles and weighted by the particle probabilities.

When the density to be approximated is Gaussian and the goal is to minimize expected mean squared error of the approximation, then Chapter 4 of [18] shows that the optimal window size is

$$h^* = \left[\frac{4}{J(d+2)} \right]^{1/(d+4)}$$

See Table 4.1 of [18]. For Σ in (3.50) we use the empirical covariance matrix of the particle distribution, which is computed as follows

$$\mu = \sum_{j=1}^{J} w^j(t) x^j(t), \quad \Sigma = \sum_{j=1}^{J} w^j(t) \left(x^j(t) - \mu\right)\left(x^j(t) - \mu\right)^T \quad (3.51)$$

For multimodal distributions, Chapter 4 of [18] recommends a using window size h somewhat smaller than $h*$.

We now turn to the question of how to perturb the state of a particle when it is a copy of a parent particle. In this case we view the parent particles as samples from a density equal to the weighted sum of Gaussian kernel densities as given in (3.49). In the perturbation algorithm given below we generate additional samples in the neighborhood of the parent particle by choosing perturbations from the kernel density K_h in (3.50) with Σ determined by (3.51). This produces a regularized particle filter. In their work on improving regularized particle filters [17], Musso, Oudjane, and Le Gland found that $h = h^* / 2$ works well for regularizing multimodal particle distributions.

Perturbation Algorithm

1. Compute the empirical covariance matrix Σ of the posterior particle distribution $\{(x^j(t_k), w^j(t_k)); j = 1, \ldots, J\}$ using (3.51).

2. For each particle $\xi^m(t_k)$ in the resampled set of particles obtained in step 4 of the resampling algorithm, make an independent draw from the kernel density K_h in (3.50) with $h = h^* / 2$ to obtain δ^m.

3. Set $\hat{\xi}^m(t) = \xi^m(t) + \delta^m$ for $t = t_k$ and $\hat{\xi}^m(t_{k'}) = \xi^m(t_k)$ for all $t_{k'} < t_k$.

If there are constraints on the target's state such as the speed constraints in the bearings-only tracking example in Chapter 1, a perturbation may put the particle state outside the constraint region. In this case one can continue sampling until a perturbation that satisfies the constraints is obtained.

If the target's state space contains components that are not part of a standard Cartesian space, then special methods must be used to perturb the target state. If for example, some of the components include target type or motion model indices, the analyst must decide how, if at all, to perturb these components.

3.3.5 Convergence

The results in [19] show that when the target state space S equals l-dimensional Cartesian space, the SIR particle filter distribution (without regularization) converges, as the number of particles $J \to \infty$, to the true posterior target state distribution. The convergence is in the following sense. Let ν_k^J denote the probability measure on S that corresponds to the particle filter measure produced at time t_k with J particles, and let ν_k denote the true posterior distribution at time t_k. Then for any bounded continuous function f defined on S

$$E\left\|\int_S f(x)dv_k^J - \int_S f(x)dv_k\right\| \to 0 \text{ as } J \to \infty$$

3.3.6 Outliers

In practice some measurements are sent to a tracker that are corrupted or are clearly outliers not generated by the target under track. These measurements need to be removed so that they are not incorporated into the tracker solution. Measurements can be checked for compatibility with the target state distribution as follows.

Let $L(y|\cdot)$ be the likelihood function for the measurement $Y = y$ at time t. Compute the likelihood $l(y|\text{target})$ of the measurement $Y = y$ given it is generated by the target at time t as follows

$$l(y|\text{target}) = \Pr\{Y = y \,|\, Y \text{ generated by target}\} = \sum_{j=1}^{J} w^j(t)L\left(y \,|\, x^j(t)\right)$$

where $\{(x^j(t), w^j(t)); j = 1,\ldots,J\}$ is the particle filter distribution at time t.

Set a threshold $\varepsilon > 0$. If $l(y) < \varepsilon$, discard the measurement. Do not incorporate it into the target state estimate.

Setting the Threshold Value ε

To determine the threshold value one might proceed as follows for the case of elliptical contacts (measurements). Let p_o be the prior probability that a measurement is an outlier. This can be estimated by examining data from past measurements produced by the sensor compared to a known target track or tracks. Define a measurement to be an outlier if the target's position at the time of the measurement is outside the n-σ ellipse of the (Gaussian) measurement uncertainty. Use the percentage of outliers as an estimate of p_o. Alternatively, p_o can be estimated subjectively.

Assuming that all values of the measurement $Y = y$ are equally likely given it is an outlier, we set

$$l(y|\text{outlier}) = \Pr\{Y = y \,|\, Y \text{ not generated by the target}\} = 1$$

Then the posterior probability $\tilde{p}_o$ of the measurement $Y = y$ being an outlier is computed as follows

$$\tilde{p}_o = \frac{p_o}{p_o + (1 - p_o)l(y|\text{tgt})} \tag{3.52}$$

If we set a threshold probability β and decide that we will discard any measurement for which $\tilde{p}_o > \beta$, then from (3.52) we see that this is equivalent to discarding any measurement for which

$$l(y \mid \text{target}) < \frac{p_o(1-\beta)}{(1-p_o)\beta} = \varepsilon \tag{3.53}$$

3.3.7 Multiple Motion Models

In some situations there may be multiple possible motion models. A target could be in transit or loitering in an area. It may have a different motion model if it is alerted to the search and seeking to avoid detection than if it is unalerted. Different target types may have different motion models. A convenient way to handle this with a particle filter is to make the motion model a (discrete) component of the target state.

Suppose there are R motion models denoted by the index $r = 1,...,R$. Consider the target state at time t to be $x'(t) = (x(t), r(t))$ where $x(t)$ is the kinematic component of the state space and $r(t)$ equals the index of the target motion model at time t. Part of the specification of the motion model is to specify the Markov process that governs the transitions from one motion model to another. The transition function can depend on only the index r or additionally on time and target state. In the case where the different motion models arise from different target types and r indexes the target type, there may be no transitions allowed.

The particle filter then proceeds as above with particle paths generated in the extended state space and with the appropriate kinematic motion (transition) model being determined by the value of the index r for each particle. As measurements are received, those particle paths most consistent with the measurements will have the highest posterior probabilities. When these particles are resampled, they will produce the largest number of split particles in the resampled particle filter distribution. This will allow the particle filter to automatically estimate the most likely motion model by computing the distribution on the motion models. If the target changes motion models, then the particle filter will detect this change as it incorporates subsequent measurements into the particle filter. This will produce a shift in the posterior distribution that moves probability to motion models that are consistent with these measurements. If the measurements provide sufficient information, the new motion model will become the high-probability one.

When perturbing the resampled particles, one usually does not perturb the component of target state corresponding to the motion model. In particular there are usually a fixed number of motion models, and the models are usually quite different from one another so there is not at natural way to perturb a motion model a "small" amount.

3.3.8 High-Dimensional State Spaces

Particle filters do not escape the "curse of dimensionality." In [20], Daum and Huang document the exponential growth in the number of particles required to obtain a good representation of the posterior target state distribution as the dimension of the target state increases. In the case of a six-dimensional target state space (e.g., three-dimensional position and velocity), one million particles are required. In [21], they propose a particle flow method that does not suffer from this exponential growth problem and does not require resampling. The latter has the benefit of allowing the computations to be easily and efficiently parallelized.

3.4 SUMMARY

This chapter defines the basic elements of Bayesian tracking, namely establishment of a prior distribution on target state and motion, employment of likelihood functions to represent measurements, and computation of posterior distributions on target state as Bayesian track estimates. When target motion is Markovian and likelihood functions depend only the target state at the time of the measurement, we obtain the Bayesian recursion for single target tracking. We show that the Kalman filtering equations are a special case of the Bayesian recursion that holds when target motion is Gaussian and measurements are linear functions of target state with additive Gaussian errors. A number of common discrete and continuous-time Gaussian motion models are presented. When these linear-Gaussian assumptions do not hold, we describe a particle filter approach for computing the Bayesian recursion and describe a particular particle filter implementation in detail. This chapter also discusses smoothing in the general case of the Bayesian tracking and presents forward-backward recursions for Kalman smoothing.

References

[1] Bar-Shalom, Y., and Fortman, T. E., *Tracking and Data Association*, New York: Academic Press, 1988.

[2] Bar-Shalom, Y., X. R. Li, and T. Kirubarajan, *Estimation with Applications to Tracking and Navigation*, New York: John Wiley & Sons, 2001.

[3] Blackman, S. S., *Multiple Target Tracking with Radar Applications*, Norwood, MA: Artech House, 1986.

[4] Blackman, S. S., and R. Popoli, *Modern Tracking Systems*, Norwood, MA: Artech House, 1999.

[5]   Hall, D. L., *Mathematical Techniques in Multisensor Data Fusion*, Norwood, MA: Artech House, 1992.

[6]   Mori, S., C-Y. Chong, C. E. Tse, and R. P. Wishner, "Tracking and classifying multiple targets without apriori identification," *IEEE Transactions on Automatic Control*, Vol. AC-31, No. 5, 1986, pp. 401-409.

[7]   Waltz, E., and J. Llinas, *Multisensor Data Fusion*, Norwood, MA: Artech House, 1990.

[8]   Ristic, B., S. Arulampalm, and N. Gordon, *Beyond the Kalman Filter*, Norwood, MA: Artech House, 2004.

[9]   Jazwinski, A. H., *Stochastic Processes and Filtering Theory*, New York: Academic Press, 1970.

[10] Castanon, D. A., B. C. Levy, and A. S. Willsky, "Algorithms for the incorporation of predictive information in surveillance theory" *International Journal of Systems Science*, Vol. 16, No. 3, 1985, pp. 367-382.

[11] Arnold, L., *Stochastic Differential Equations: Theory and Applications*, Mineola: NY, Dover Books on Mathematics, 2013.

[12] Rauch, H.E., F. Tung, and C. T. Striebel, "Maximum likelihood estimates of linear dynamic systems" *American Institute of Aeronautics and Astronautics Journal* Vol 3, August 1965, pp. 1445–1450.

[13] Anderson, B. D. O., and J. B. Moore, *Optimal Filtering*, Mineola, NY: Dover, 2005.

[14] Gelb, A (ed.), Applied Optimal Estimation, Cambridge, MA: The Analytic Sciences Corporation, 1974.

[15] DeGroot, M. H., *Optimal Statistical Decisions*, New York: Wiley Classics Library, 2004.

[16] Doucet, A., N. de Freita, and N. Gordon (eds.), *Sequential Monte Carlo Methods in Practice*, New York: Springer, 2001.

[17] Muso, M., N. Oudjane, and F. Le Gland, "Improving regularized particle filters" in *Sequential Monte Carlo Methods in Practice* (Doucet, A., N. de Freitas, and N. Gordon (eds.)), New York: Springer, 2001.

[18] Silverman, B. W., *Density Estimation for Statistics and Data Analysis*, New York: Chapman Hall, 1986.

[19] Crisan, D, "Particle Filters—A theoretical perspective" in *Sequential Monte Carlo Methods in Practice* (Doucet, A., N. de Freitas, and N. Gordon (eds.)), New York: Springer, 2001.

[20] Daum, F., and J. Huang, "Particle degeneracy: root cause and solution," *Proceedings of SPIE* Vol. 8050, 2011.

[21] Daum, F., and J. Huang, "Particle flow with non-zero diffusion for nonlinear filters," *Proceedings of SPIE*, Vol. 8745, 2013.

Chapter 4

Classical Multiple Target Tracking

Chapter 3 presents the Bayesian approach to tracking a single target. Central to this approach is the concept of using likelihood functions defined on the target state space to represent the information from diverse sensors. Likelihood functions provide a common currency for valuing and combining information from disparate sensors, particularly when these sensors have different measurement spaces.

There are two key assumptions made in Chapter 3. First, there is one and only one target present, and second, all observations are generated by this target. For multiple target tracking, we allow the possibility that there is more than one target present or even that there are no targets present. The sensor responses (contacts) may be due to any one of the targets or to none of them. In the latter case, we call the sensor response a false measurement. In this case the observation may be due to a number of causes not related to a target. The observation could be the result of statistical noise or environmentally caused signals such as clutter.

The chief difficulty in multiple target tracking is in deciding which target generated each sensor response or whether the response is a false measurement. The process of assigning a sensor response or contact to a target is called data association. If the association of contacts to targets were always unambiguous, we could decompose the multiple target tracking problem into independent single target problems and proceed as in Chapter 3. When association is ambiguous, the tracking and association problems become coupled and difficult to solve.

This chapter develops classical, multiple target tracking from the point of view of Bayesian inference. In classical multiple target tracking, the problem is divided into two steps, association and estimation. Step 1 associates contacts with targets. Step 2 uses the contacts associated with each target to produce an estimate of that target's state. Complications arise when there is more than one reasonable way to associate contacts with targets. The classical approach to this problem is to form association hypotheses and to use multiple hypothesis tracking (MHT) which is described in this chapter. In this approach alternative hypotheses are formed to explain the source of the observations. Each hypothesis assigns observations to targets or false measurements. For each hypothesis, MHT computes the probability that the hypothesis is correct. This is also the probability that the target state estimates that result from this hypothesis are correct. The posterior distribution on system state (number of targets and the state of each target) is a mixture of distributions. To see this, note that for each association hypothesis, there is a probability distribution on the joint state of the targets identified in the hypothesis. The Bayesian posterior is a mixture of these joint multitarget distributions weighted by the probability that the corresponding association hypothesis is the correct one.

A difficulty with the MHT approach is that the number of association hypotheses grows exponentially as time and the number of contacts increase. MHT algorithms often deal with this problem by limiting themselves to a fixed number of the highest probability hypotheses. In addition, most MHT algorithms display the target state distributions for only the highest probability association hypothesis.

Multiple target tracking algorithms that are capable of handling large numbers of targets and contacts must approximate the problem in some fashion. Limiting the number of association hypotheses in MHT is one approach. Two other approaches will be presented in this chapter, joint probabilistic data association (JPDA) and probabilistic multiple hypothesis tracking (PMHT).

MHT makes the assumption that contacts arrive in scans. In a scan each target generates at most one contact, and each contact is due to at most one target. Thus a contact is associated to at most one target and this association, although unknown, is hard (i.e., a contact is or is not associated to a target). There is no in between. Soft association relaxes this assumption by allowing for partial or probabilistic associations. JPDA and PMHT are examples of soft association algorithms. In the case of PMHT, contacts arrive in scans but PMHT allows for possibility that more than one contact in a scan can be associated to a single target. Both the JPDA and PMHT approximations produce simpler algorithms than MHT but at the cost of reduced accuracy.

There are numerous books and articles on multiple target tracking that deal with the many variations and approaches to this problem. Many of these discuss the practical aspects of implementing multiple target trackers and compare approaches. See for example [1–12]. A number of these references deal extensively with the linear-Gaussian case. We do not intend to cover that material in detail. The goal of this chapter is to develop the basic features of classical multiple target tracking using a rigorous Bayesian inference framework and to describe three algorithms that approximate the computations required by the Bayesian model.

Section 4.1 describes the general model that we use for the multiple target tracking problem. Section 4.1.3 presents the Bayesian recursion for multiple target tracking. This is a very general recursion that applies to problems that are nonlinear and non-Gaussian as well as to standard linear-Gaussian situations. This general model does not require the notions of contact and association. The computations required to implement this general approach can be daunting, but [13, 14] have developed particle filter methods to do this.

Section 4.2 defines the important special case of MHT along with the notions associated with it, namely scans, associations, association hypotheses, association probabilities, conditional target state distributions, and the MHT decomposition.

MHT is most useful when the conditional target distributions (conditioned on an association hypothesis) are independent of one another. In Section 4.3, we find conditions under which the independence of the target distributions is guaranteed and present the MHT recursion for this case. Section 4.4 gives a specialized version of the MHT recursion that holds when independence is coupled with linear-Gaussian assumptions. This is the version often used by MHT algorithms.

Section 4.5 describes a nonlinear version of JPDA developed in [2] and presents a particle filer implementation of this algorithm. Section 4.6 presents the PMHT algorithm developed in [15] along with a nonlinear extension of it.

Chapter 5 considers multiple target approximations that do not require the explicit association of contacts to targets. Chapter 6 considers multiple target tracking using tracker-generated measurments.

4.1 MULTIPLE TARGET TRACKING

This section introduces the classical, nonlinear, multiple target tracking problem and extends the Bayesian recursion for single target tracking to multiple targets. As in Chapter 3, we employ a continuous-discrete formulation of tracking where the target motion takes place in continuous time, but observations are received at a discrete sequence $0 \leq t_1 \leq \ldots \leq t_K$ of possibly random times.

4.1.1 Multiple Target Motion Model

Section 3.1.1 represents the prior knowledge about the single target's state and its motion through the target state space S in terms of a stochastic process $\{X(t); t \geq 0\}$, where $X(t)$ is the target state at time t. For multiple targets we generalize this notion as follows.

We begin the multiple target tracking problem at time $t = 0$. The total number of targets is unknown but bounded by $\bar{N}$, which is known. We assume a known bound on the number of targets, because it allows us to simplify the presentation and produces no restriction in practice.

We add an additional state ϕ to the target state space S. If a target is not present, we say that it is in state ϕ. Let $S^+ = S \cup \{\phi\}$ be the augmented state space for a single target and $\mathbf{S}^+ = S^+ \times \cdots \times S^+$ be the joint target state space where the product is taken $\bar{N}$ times.

Multiple Target Motion Process

Our prior knowledge about the targets and their "movements" through the state space $\mathbf{S}^+$ is expressed as a stochastic process $\mathbf{X} = \{\mathbf{X}(t) \; ; \; t \geq 0\}$. Specifically, let $\mathbf{X}(t) = (X_1(t), \ldots, X_{\bar{N}}(t))$ be the state of the system at time t where $X_n(t) \in S^+$ is the state of target n at time t. We use the term "state of the system" to mean the joint state of all the targets. The value of the random variable $X_n(t)$ indicates whether target n is present and what state it is in if present; $X_n(t) = \phi$ means that target n is not present at time t. The number of components of $\mathbf{X}(t)$ with states not equal to ϕ at time t gives the number of targets present at time t. We assume that the stochastic process $\mathbf{X}$ is Markovian in the state space $\mathbf{S}^+$ and that the process has an associated transition function. Let

$$q_k(\mathbf{s}_k \mid \mathbf{s}_{k-1}) = \Pr\{\mathbf{X}(t_k) = \mathbf{s}_k \mid \mathbf{X}(t_{k-1}) = \mathbf{s}_{k-1}\} \quad \text{for } k \geq 1$$

and let q_0 be the probability (density) function for $\mathbf{X}(0)$. By the Markov assumption

$$q(\mathbf{s}_1, \ldots, \mathbf{s}_K) \equiv \Pr\{\mathbf{X}(t_1) = \mathbf{s}_1, \ldots, \mathbf{X}(t_K) = \mathbf{s}_K\} = \int \prod_{k=1}^{K} q_k(\mathbf{s}_k \mid \mathbf{s}_{k-1}) q_0(\mathbf{s}_0) \, d\mathbf{s}_0 \quad (4.1)$$

The motion model can allow for targets to arrive (transition from ϕ to S) and depart (transition from S to ϕ) as time progresses.

Integration

The symbol $d\mathbf{s}$ will be used to indicate integration with respect to the measure on $\mathbf{S}^+$ whether it is discrete or not. When the measure is discrete, integrals become summations. Similarly, when we use the notation Pr we shall mean either probability or probability density as appropriate.

4.1.2 Multiple Target Likelihood Functions

We suppose that there is a set of sensors that report observations at a discrete sequence of possibly random times. These sensors may be of different types and report different information. The sensors may report only when they have a contact or on a regular basis. Let $Y(t, j)$ be an observation (measurement) from sensor j at time t. Observations from sensor j take values in the measurement space $\mathcal{M}_j$. Each sensor may have a different measurement space.

We assume that for each sensor j, we can compute

$$l(t, j, y \mid \mathbf{s}) = \Pr\{Y(t, j) = y \mid \mathbf{X}(t) = \mathbf{s}\} \text{ for } y \in \mathcal{M}_j \text{ and } \mathbf{s} \in \mathbf{S}^+ \qquad (4.2)$$

The assumption that we can compute the probabilities in (4.2) means that we know the probability distribution of the sensor response conditioned on system state $\mathbf{s}$. In contrast to Chapter 3, the likelihood functions in this chapter can depend on the joint state of all the targets. The relationship between the observation and the state $\mathbf{s}$ may be linear or nonlinear and the probability distribution may be Gaussian or non-Gaussian.

Suppose that by time t, we have obtained observations at the set of discrete times $0 \le t_1 \le \ldots \le t_K \le t$. To allow for the possibility that we may receive more than one measurement at a given time, we let Y_k be the set of measurements received at time t_k. Let y_k denote a value of the random variable Y_k. We extend (4.2) to assume that we can compute

$$L_k(y_k \mid \mathbf{s}) = \Pr\{Y_k = y_k \mid \mathbf{X}(t_k) = \mathbf{s}\} \text{ for } \mathbf{s} \in \mathbf{S}^+ \qquad (4.3)$$

We call $L_k(y_k \mid \cdot)$ the *likelihood function* for the observation $Y_k = y_k$. The computation in (4.3) can account for correlation among sensor responses if that is required.

Let $\mathbf{Y}(K) = (Y_1, Y_2, \ldots, Y_K)$ and $\mathbf{y}_K = (y_1, \ldots, y_K)$. Define

$$L(\mathbf{y}_K \mid \mathbf{s}_1, \ldots, \mathbf{s}_K) = \Pr\{\mathbf{Y}(K) = \mathbf{y}_K \mid X(t_1) = \mathbf{s}_1, \ldots, X(t_K) = \mathbf{s}_K\}$$

In parallel with Chapter 3, we assume that

$$\Pr\{\mathbf{Y}(K) = \mathbf{y}_K \mid \mathbf{X}(u) = \mathbf{s}(u), 0 \le u \le t\} = L(\mathbf{y}_K \mid \mathbf{s}(t_1), \ldots, \mathbf{s}(t_K)) \qquad (4.4)$$

and

$$L(\mathbf{y}_K \mid \mathbf{s}_1, \ldots, \mathbf{s}_K) = \prod_{k=1}^{K} L_k(y_k \mid \mathbf{s}_k) \qquad (4.5)$$

In (4.4) we are assuming that the distribution of the sensor responses at the times $\{t_1, \ldots, t_K\}$ depends only on the system states at those times. In (4.5), we are assuming independence of the sensor response distributions across the observation times. The effect of both assumptions is to assume that the sensor response at time t_k depends only on the system state at that time.

4.1.3 Bayesian Recursion for Multiple Targets

In parallel with the Bayesian recursion for single target tracking, we can state the Bayesian recursion for multiple target tracking.

The posterior distribution on the multiple target state is given by

$$
p(t_K, \mathbf{s}_K) = \frac{\Pr\{\mathbf{Y}(K) = \mathbf{y}_K \text{ and } \mathbf{X}(t_k) = \mathbf{s}_K\}}{\Pr\{\mathbf{Y}(K) = \mathbf{y}_K\}}
$$

$$
= \frac{\int L(\mathbf{y}_K \mid \mathbf{s}_1, \ldots, \mathbf{s}_K) q(\mathbf{s}_1, \ldots, \mathbf{s}_K) d\mathbf{s}_1 \cdots d\mathbf{s}_{K-1}}{\int L(\mathbf{y}_K \mid \mathbf{s}_1, \ldots, \mathbf{s}_K) q(\mathbf{s}_1, \ldots, \mathbf{s}_K) d\mathbf{s}_1 \cdots d\mathbf{s}_K}
$$

By (4.1) and (4.5) we have

$$
p(t_K, \mathbf{s}_K) = \frac{1}{C'} \int \prod_{k=1}^{K} L_k(y_k \mid \mathbf{s}_k) \prod_{k=1}^{K} q_k(\mathbf{s}_k \mid \mathbf{s}_{k-1}) q_0(\mathbf{s}_0) d\mathbf{s}_0 \cdots d\mathbf{s}_{K-1}
$$

$$
= \frac{1}{C'} L_K(y_K \mid \mathbf{s}_K) \int q_K(\mathbf{s}_K \mid \mathbf{s}_{K-1})
$$

$$
\times \left[\int \prod_{k=1}^{K-1} L_k(y_k \mid \mathbf{s}_k) q_k(\mathbf{s}_k \mid \mathbf{s}_{k-1}) q_0(\mathbf{s}_0) d\mathbf{s}_0 \cdots d\mathbf{s}_{K-2} \right] d\mathbf{s}_{K-1}
$$

and

$$
p(t_K, \mathbf{s}_K) = \frac{1}{C} L_K(y_K \mid \mathbf{s}_K) \int q_K(\mathbf{s}_K \mid \mathbf{s}_{K-1}) \, p(t_{K-1}, \mathbf{s}_{K-1}) d\mathbf{s}_{K-1}
$$

where C' and C are constants that normalize the right-hand sides of the above equations into probability distributions.

We can now state the Bayesian recursion for multiple target tracking. The form of this recursion is completely parallel to the single target recursion in Chapter 3.

Bayesian Recursion for Multiple Target Tracking

Initial distribution $\qquad p(t_0, \mathbf{s}_0) = q_0(\mathbf{s}_0)$ for $\mathbf{s}_0 \in \mathbf{S}^+$ $\qquad$ (4.6)

For $k \geq 1$ and $\mathbf{s}_k \in \mathbf{S}^+$,

Motion update $\qquad p^-(t_k, \mathbf{s}_k) = \int q_k(\mathbf{s}_k \mid \mathbf{s}_{k-1}) p(t_{k-1}, \mathbf{s}_{k-1}) d\mathbf{s}_{k-1}$ $\qquad$ (4.7)

Measurement likelihood $\qquad L_k(y_k \mid \mathbf{s}_k) = \Pr\{Y_k = y_k \mid \mathbf{X}(t_k) = \mathbf{s}_k\}$ $\qquad$ (4.8)

Information update $\qquad p(t_k, \mathbf{s}_k) = \dfrac{1}{C} L_k(y_k \mid \mathbf{s}_k) p^-(t_k, \mathbf{s}_k)$ $\qquad$ (4.9)

4.2 MULTIPLE HYPOTHESIS TRACKING

In this section, we develop the MHT approach to multiple target tracking. In this approach one considers (at least conceptually) all possible hypotheses for associating measurements to targets. For each hypothesis, MHT computes the joint target state distribution conditioned on that hypothesis. The posterior distribution on system state is then obtained as a mixture of these conditional distributions weighted by the association probabilities (i.e., the probability that the hypothesis conditioned upon is the correct one). This produces the MHT decomposition given in Section 4.2.5. However, this decomposition will be of limited usefulness unless the conditional target state distributions are independent. Section 4.3 provides conditions under which independence holds.

We begin by defining contacts, scans, and association hypotheses. Often the response of a sensor is thresholded in some manner to determine when a detection is called. For example, the signal-to-noise ratio (SNR) may be estimated and a detection called when SNR exceeds a threshold. Once the detection is called, the sensor response may be analyzed to produce a measurement, for example, a (partial) estimate of target state such as position or bearing.

4.2.1 Contacts

Definition 4.1. A *contact* is an observation that consists of a called detection and a measurement.

For the remainder of this chapter we assume that all observations are in the form of contacts. By limiting the sensor responses to contacts, we are restricting ourselves to responses in which the signal level of the target, as seen at the sensor, is high enough to call a contact. In order to perform MHT we will further assume that contacts come in the form of scans as defined in Section 4.2.2. These restrictions are necessary for MHT. However, we shall see in Chapters 5–7 that one can (and in some cases must) perform tracking without these assumptions being satisfied.

Elliptical Contacts

The measurement Θ_k associated with a contact at time t_k often has the form

$$\Theta_k = HX(t_k) + \varepsilon_k$$

where

> Θ_k is an r-dimensional real column vector
>
> $X(t_k)$ is an l-dimensional real column vector
>
> H is an $r \times l$ matrix
>
> ε_k is r-dimensional Gaussian with mean 0 and covariance R_k

Suppose that $r = l = 2$ so that the target state space is the plane of points (x_1, x_2) and that $H = \mathbf{I}$, the two-dimensional identity matrix. The likelihood function for this measurement is the two-dimensional Gaussian density function given in Section 2.3.1. The contact and its measurement uncertainty are often represented by an ellipse as shown in Figure 4.1. We call this an *elliptical contact*. The center of the ellipse is equal to the measurement $\Theta_k = \theta_k$. The ellipse is typically the set of points $x = (x_1, x_2)$ that satisfy

$$\left(x - \theta_k\right)^T R_k^{-1}\left(x - \theta_k\right) = 4$$

This is the ellipse of points at "distance" 2 units from Y_k measured in the normalized distance units determined by the covariance matrix R_k. This ellipse is often called the 2-σ ellipse and is used to graphically represent a contact with a bivariate Gaussian measurement error. In this two dimensional case, there is an 86% probability of the target being in the 2-σ ellipse.

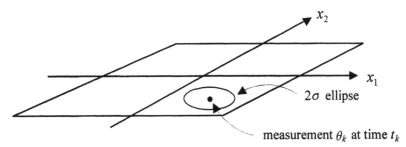

Figure 4.1 An elliptical contact.

4.2.2 Scans

We now further limit the class of measurements to scans.

Definition 4.2. The set of contacts Y_k at time t_k is a *scan* if it consists of a set C_k of contacts such that the measurement from each contact is associated with at most one target and each target generates at most one contact.

Some of these contacts may contain false measurements, and some targets may not be detected on a given scan.

More than one sensor group can report a scan at the same time. In this case, the contact reports from each sensor group are treated as separate scans with the same reporting time. As a result, we can have $t_k = t_{k+1}$. A scan can also consist of a single contact report. Figure 4.2 shows a scan that consists of multiple elliptical contacts.

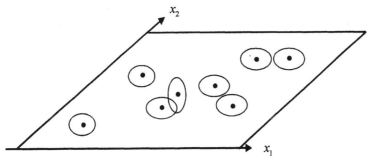

Figure 4.2 A scan of multiple elliptical contacts.

4.2.3 Data Association Hypotheses

A data association hypothesis, h, is defined as follows. Let

$$C_j = \text{ set of contacts in the } j\text{th scan}$$

$$C(1:k) = \text{ set of all contacts reported in the first } k \text{ scans}$$

$$= \bigcup_{j=1}^{k} C_j$$

Definition 4.3. A data association hypothesis h on $C(1:k)$ is a mapping

$$h : C(1:k) \rightarrow \{0, 1, \ldots, \bar{N}\} \text{ such that}$$

$h(c) = n > 0$ means the measurement from contact c is associated to target n,

$h(c) = 0$ means contact c is associated to a false measurement, and

no target has more than one measurement from a scan associated to it.

Figure 4.3 shows a data association hypothesis for $k = 3$ scans. A line connecting a series of contacts indicates that those contacts are associated to the same target in this hypothesis. One can see how the association hypothesis partitions the data into sets of contacts associated with each target and with false measurements. For example, the dashed line indicating target 3 in Figure 4.3 identifies a contact in scan 1 and a contact in scan 3 as being associated to target 3. Since missed detections are possible, there may be targets (such as target 3 in scan 2) that have no contacts associated to them in a scan. Target 7 is a "new" target at time 2 and target 8 is new at time 3.

Let

$$H(k) = \text{ set of all data association hypotheses on } C(1:k)$$

A hypothesis $h \in H(k)$ partitions $C(1:k)$ into subsets

$$T_k(n) = \{c \in C(1:k) : h(c) = n\} \text{ for } n = 0, 1, \ldots, \bar{N}$$

where $T_k(n)$ is the subset of·contacts associated to target n for $n > 0$ and $T_k(0)$ is the subset of contacts associated to false measurements.

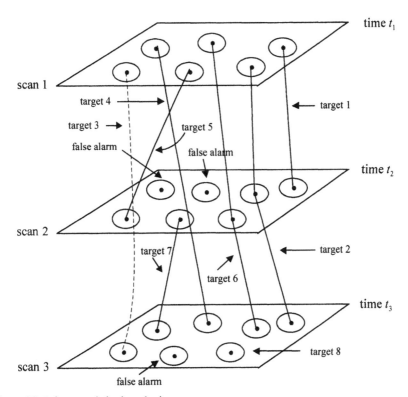

Figure 4.3 A data association hypothesis.

4.2.4 Scans and Scan Association Hypotheses

Let us turn our attention to data association hypotheses for a single scan. For the kth scan Y_k, let M_k = number of contacts in scan Y_k.

Definition 4.4. A function $\gamma : \{1,...,M_k\} \to \{0,...,\bar{N}\}$ is a *scan association hypothesis* if

$\gamma(m) = n > 0$ means contact m is associated to target n,
$\gamma(m) = 0$ means contact m is associated to a false measurement, and
no two contacts are assigned to the same positive number (target).

Let

$$\Gamma_k = \text{the set of all scan association hypotheses on the scan } Y_k$$

Figure 4.4 shows a scan association hypothesis γ. In this hypothesis, contacts 1 and 2 are associated to false measurements. Contact 6 is associated to target 5.

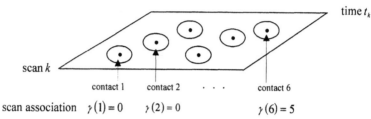

Figure 4.4 A scan association hypothesis, γ.

Consider a data association hypothesis $h_K \in H(K)$. We may think of h_K as composed of K scan association hypotheses $\{\gamma_1,...,\gamma_K\}$ where γ_k is the association hypothesis for the kth scan of contacts. The hypothesis h_K is the extension of the hypothesis $h_{K-1} = \{\gamma_1,...,\gamma_{K-1}\} \in H(K-1)$. That is, h_K is composed of h_{K-1} plus γ_K. We write this as $h_K = h_{K-1} \wedge \gamma_K$.

4.2.4.1 Scan Association Likelihood Function

We assume that for each scan Y_k, we can calculate the likelihood of obtaining the measurements (contacts) in the scan given a scan association hypothesis γ and system state s_k. Specifically we assume that we can calculate the scan association likelihood function

$$\ell_k(y_k \mid s_k \wedge \gamma) = \Pr\{Y_k = y_k \mid \gamma \wedge (X(t_k) = s_k)\} \text{ for } s_k \in S^+ \text{ and } \gamma \in \Gamma_k \quad (4.10)$$

The conditioning on the right-hand side of (4.10) means that we are conditioning on the scan association hypothesis γ as well as the system state s_k. For clarity we have put parentheses around $X(t_k) = s_k$ in (4.10), but in the future we will use parentheses only when confusion is possible.

A scan association hypothesis γ specifies which contacts are associated to which targets and the set of contacts that are associated to false measurements. Recall that a system state s_k specifies which targets are present and the states in S of the targets that are present. As a function of s_k, the likelihood of the scan data computed in (4.10) will account for the probability of detecting the targets to which contacts are associated as well as failing to detect the remaining targets present. In addition it will account for the probability of the set of contacts associated to false measurements. An example of a scan association likelihood function is given in Section 4.3.2 when we consider independent MHT.

The likelihood function for the scan $Y_k = y_k$ is

$$
\begin{aligned}
L_k\left(y_k \mid \mathbf{s}_k\right) &= \Pr\left\{Y_k = y_k \mid \mathbf{X}(t_k) = \mathbf{s}_k\right\} \\
&= \sum_{\gamma \in \Gamma_k} \Pr\left\{Y_k = y_k \mid \gamma \wedge \mathbf{X}(t_k) = \mathbf{s}_k\right\} \Pr\left\{\gamma \mid \mathbf{X}(t_k) = \mathbf{s}_k\right\} \\
&= \sum_{\gamma \in \Gamma_k} \ell_k\left(y_k \mid \gamma \wedge \mathbf{s}_k\right) \Pr\left\{\gamma \mid \mathbf{X}(t_k) = \mathbf{s}_k\right\} \text{ for } \mathbf{s}_k \in \mathbf{S}^+
\end{aligned}
$$

We will assume that the (prior) probability of a scan association does not depend on system state so that

$$
\Pr\left\{\gamma \mid \mathbf{X}(t_k) = \mathbf{s}_k\right\} = \Pr\left\{\gamma\right\} \text{ for } \mathbf{s}_k \in \mathbf{S}^+ \text{ and } \gamma \in \Gamma_k
$$

As a result

$$
L_k\left(y_k \mid \mathbf{s}_k\right) = \sum_{\gamma \in \Gamma_k} \ell_k\left(y_k \mid \gamma \wedge \mathbf{s}_k\right) \Pr\left\{\gamma\right\} \text{ for } \mathbf{s}_k \in \mathbf{S}^+ \tag{4.11}
$$

4.2.4.2 Data Association Likelihood Function

Recall that $\mathbf{Y}(K) = \mathbf{y}_K$ is the set of observations (measurements) contained in the first K scans and $H(K)$ is the set of data association hypotheses defined on these scans. From (4.5) it follows that conditioned on $h \in H(K)$, the likelihood of the measurements received at times $t_1, \ldots, t_K$ depends only on the system state values at those times. Specifically

$$
\begin{aligned}
\Pr\left\{\mathbf{Y}(K) = \mathbf{y}_K \mid h \wedge \mathbf{X}(u) = \mathbf{x}_u; \ 0 \leq u \leq t_K\right\} \\
= \Pr\left\{\mathbf{Y}(K) = \mathbf{y}_K \mid h \wedge \mathbf{X}(t_k) = \mathbf{s}_k; \ k = 1, \ldots, K\right\}
\end{aligned} \tag{4.12}
$$

We define the data association likelihood function l as

$$
l\left(\mathbf{y}_K \mid h \wedge (\mathbf{s}_1, \ldots, \mathbf{s}_K)\right) = \Pr\left\{\mathbf{Y}(K) = \mathbf{y}_K \mid h \wedge \mathbf{X}(t_k) = \mathbf{s}_k; \ k = 1, \ldots, K\right\}
$$

In addition, we assume that the scan association likelihoods are independent given $h \wedge (\mathbf{s}_1, \ldots, \mathbf{s}_K)$, in other words,

$$
l\left(\mathbf{y}_K \mid h \wedge (\mathbf{s}_1, \ldots, \mathbf{s}_K)\right) = \prod_{k=1}^{K} \ell_k\left(y_k \mid \gamma_k \wedge \mathbf{s}_k\right) \tag{4.13}
$$

where $\mathbf{y}_K = (y_1, \ldots y_K)$ and $h = \{\gamma_1, \ldots, \gamma_K\}$.

Finally we assume that the prior probability of data association hypothesis h is equal to the product of the prior probabilities of its constituent scan association hypotheses. Specifically,

$$\Pr\{h\} = \prod_{k=1}^{K} \Pr\{\gamma_k\} \text{ where } h = \{\gamma_1,\ldots,\gamma_K\} \qquad (4.14)$$

4.2.5 Multiple Hypothesis Tracking Decomposition

Conceptually, MHT proceeds as follows: It calculates the posterior distribution on the system state at time t_K (given that data association hypothesis h is true) and the probability $\alpha(h\,|\,\mathbf{y}_K)$ that hypothesis h is true given $\mathbf{Y}(K) = \mathbf{y}_K$ for each $h \in H(K)$. That is, for each $h \in H(K)$, it computes

$$p(t_K, \mathbf{s}_K \mid h) = \Pr\{\mathbf{X}(t_K) = \mathbf{s}_K \mid h \wedge \mathbf{Y}(K) = \mathbf{y}_K\} \qquad (4.15)$$

and

$$\alpha(h\,|\,\mathbf{y}_K) = \Pr\{h\,|\,\mathbf{Y}(K) = \mathbf{y}_K\} = \frac{\Pr\{h \wedge \mathbf{Y}(K) = \mathbf{y}_K\}}{\Pr\{\mathbf{Y}(K) = \mathbf{y}_K\}} \qquad (4.16)$$

Having done this, one can compute the Bayesian posterior on system state by

$$p(t_K, \mathbf{s}_K) = \sum_{h \in H(K)} \alpha(h\,|\,\mathbf{y}_K)\, p(t_K, \mathbf{s}_K \mid h) \qquad (4.17)$$

This is the MHT decomposition. Section 4.3 shows how to compute $p(t_K, \mathbf{s}_K \mid h)$ and $\alpha(h\,|\,\mathbf{y}_K)$ in a joint recursion when certain independence assumptions hold.

4.2.5.1 Conditional Target Distributions

We call the distributions conditioned on the truth of an hypothesis *conditional target distributions*. The distribution $p(t_K, \cdot \mid h)$ in (4.15) is an example of a conditional joint target state distribution. These distributions are always conditioned on the data received, specifically, $\mathbf{Y}(K) = \mathbf{y}_K$, but we do not show this conditioning in our notation, $p(t_K, \mathbf{s}_K \mid h)$.

Let $h_K = \{\gamma_1,\ldots,\gamma_K\}$, then

$$\begin{aligned}
p(t_K, \mathbf{s}_K \mid h_K) &= \Pr\{\mathbf{X}(t_K) = \mathbf{s}_K \mid h_K \wedge \mathbf{Y}(K) = \mathbf{y}_K\} \\
&= \frac{\Pr\{(\mathbf{X}(t_K) = \mathbf{s}_K) \wedge h_K \wedge (\mathbf{Y}(K) = \mathbf{y}_K)\}}{\Pr\{h_K \wedge \mathbf{Y}(K) = \mathbf{y}_K\}}
\end{aligned}$$

and by (4.1) and the data association likelihood function in (4.13), we have

$$\Pr\left\{(\mathbf{X}(t_K) = \mathbf{s}_K) \wedge h_K \wedge (\mathbf{Y}(K) = \mathbf{y}_K)\right\}$$
$$= \Pr\{h_K\} \int \left\{\prod_{k=1}^{K} \ell_k(y_k \mid \gamma_k \wedge \mathbf{s}_k) \prod_{k=1}^{K} q_k(\mathbf{s}_k \mid \mathbf{s}_{k-1}) q_0(\mathbf{s}_0)\right\} d\mathbf{s}_0 d\mathbf{s}_1 \cdots d\mathbf{s}_{K-1} \tag{4.18}$$

Using (4.14), we can write

$$p(t_K, \mathbf{s}_K \mid h_K)$$
$$= \frac{1}{C(h_K)} \int \left\{\prod_{k=1}^{K} \ell_k(y_k \mid \gamma_k \wedge \mathbf{s}_k) \Pr\{\gamma_k\} q_k(\mathbf{s}_k \mid \mathbf{s}_{k-1})\right\} q_0(\mathbf{s}_0) d\mathbf{s}_0 d\mathbf{s}_1 \cdots d\mathbf{s}_{K-1} \tag{4.19}$$

where $C(h_K)$ is the normalizing factor that makes $p(t_K, \cdot \mid h_K)$ a probability distribution. Of course

$$C(h_K) = \int \left\{\prod_{k=1}^{K} \ell_k(y_k \mid \gamma_k \wedge \mathbf{s}_k) \Pr\{\gamma_k\} q_k(\mathbf{s}_k \mid \mathbf{s}_{k-1})\right\} q_0(\mathbf{s}_0) d\mathbf{s}_0 d\mathbf{s}_1 \cdots d\mathbf{s}_K$$
$$= \Pr\left\{h_K \wedge \mathbf{Y}(K) = \mathbf{y}_K\right\} \tag{4.20}$$

Recall that $h_K = h_{K-1} \wedge \gamma_K$, and let

$$p^-(t_K, \mathbf{s}_K \mid h_{K-1}) = \int q_K(\mathbf{s}_K \mid \mathbf{s}_{K-1}) p(t_{K-1}, \mathbf{s}_{K-1} \mid h_{K-1}) d\mathbf{s}_{K-1}$$

Note that $p^-(t_K, \cdot \mid h_{K-1})$ is the distribution on system state updated for motion to time t_K but conditioned only on the observations up to time t_{K-1} and the association hypothesis h_{K-1}. From this definition and (4.19), we have

$$p(t_K, \mathbf{s}_K \mid h_K)$$
$$= \frac{C(h_{K-1})}{C(h_K)} \ell_K(y_K \mid \gamma_K \wedge \mathbf{s}_k) \Pr\{\gamma_K\} \int q_K(\mathbf{s}_K \mid \mathbf{s}_{K-1}) p(t_{K-1}, \mathbf{s}_{K-1} \mid h_{K-1}) d\mathbf{s}_{K-1}$$
$$= \frac{C(h_{K-1})}{C(h_K)} \ell_K(y_K \mid \gamma_K \wedge \mathbf{s}_k) \Pr\{\gamma_K\} p^-(t_K, \mathbf{s}_K \mid h_{K-1}) \tag{4.21}$$
$$= \frac{\ell_K(y_K \mid \gamma_K \wedge \mathbf{s}_k) p^-(t_K, \mathbf{s}_K \mid h_{K-1})}{\int \ell_K(y_K \mid \gamma_K \wedge \mathbf{s}_k) p^-(t_K, \mathbf{s}_K \mid h_{K-1}) d\mathbf{s}_K}$$

The last equality in (4.21) holds because $p(t_K, \cdot | h_K)$ is a probability (density) function on $\mathbf{S}^+$. In a similar manner, one can show that

$$C(h_K) = C(h_{K-1}) \Pr\{\gamma_K\} \int \ell_K (y_K \mid \gamma_K \wedge \mathbf{s}_K) p^- (t_K, \mathbf{s}_K \mid h_{K-1}) d\mathbf{s}_K \qquad (4.22)$$

4.2.5.2 Association Probabilities

We can now calculate the association probability $\alpha(h \mid \mathbf{y}_K)$ as follows.

$$
\begin{aligned}
\alpha(h \mid \mathbf{y}_K) &= \Pr\{h \mid \mathbf{Y}(K) = \mathbf{y}_K\} \\
&= \frac{\Pr\{h \wedge \mathbf{Y}(K) = \mathbf{y}_K\}}{\Pr\{\mathbf{Y}(K) = \mathbf{y}_K\}} = \frac{C(h)}{\displaystyle\sum_{h' \in H(K)} C(h')} \quad \text{for } h \in H(K) \qquad (4.23)
\end{aligned}
$$

where the last equality in (4.23) follows from (4.20).

4.3 INDEPENDENT MULTIPLE HYPOTHESIS TRACKING

It is difficult to compute and maintain joint target state distributions except for small numbers of targets because the size of the joint target state space becomes overwhelmingly large. Thus, while (4.17) presents a theoretical solution to the multiple target tracking problem, it will not generally be useful in applications unless the target state distributions (conditioned on a data association hypothesis) are independent. The remainder of this section develops the conditions under which these conditional target state distributions are independent and then derives the recursion for independent MHT.

In this section we introduce the class of conditionally independent scan association likelihood functions and show that when the scan association likelihood functions are conditionally independent and the target motion models are independent, then the target state distributions are independent conditioned on a data association hypothesis. When this is true, we do not have to maintain a joint target state space distribution provided we condition on a data association hypothesis. In this case $p(t_K, \mathbf{s}_K \mid h)$ in (4.17) is equal to the product of independent probability distributions on the $\bar{N}$ possible targets, and the MHT decomposition in (4.17) becomes

$$p(t_K, \mathbf{s}_K) = \sum_{h \in H(K)} \alpha(h \mid \mathbf{y}_K) p(t_K, \mathbf{s}_K \mid h)$$

$$= \sum_{h \in H(K)} \alpha(h \mid \mathbf{y}_K) \prod_{n=1}^{\bar{N}} p_n(t_K, x_n \mid h) \text{ for } \mathbf{s}_K = (x_1, \ldots, x_{\bar{N}}) \in \mathbf{S}^+$$

Note that if a target is not present in a given hypothesis at time t_k, then $\mathrm{Pr}\{\phi\} = 1$ for that target.

4.3.1 Conditionally Independent Association Likelihoods

Prior to this section we have made no special assumptions about the scan association likelihood function

$$\ell_k\left(y_k \mid \mathbf{s}_k \wedge \gamma\right) = \mathbf{Pr}\left\{Y_k = y_k \mid \mathbf{X}(t_k) = \mathbf{s}_k \wedge \gamma\right\} \text{ for } \mathbf{s}_k \in \mathbf{S}^+$$

In many cases, however, the likelihood of a scan of contacts given a scan association hypothesis satisfies an independence assumption when conditioned on system state.

Definition 4.5. The likelihood of a scan $Y_k = y_k$ obtained at time t_k is *conditionally independent* if and only if for all scan association hypotheses $\gamma \in \Gamma_k$,

$$\ell_k\left(y_k \mid \mathbf{s}_k = (x_1, \ldots, x_{\bar{N}}) \wedge \gamma\right) = \mathbf{Pr}\left\{Y_k = y_k \mid \mathbf{X}(t_k) = (x_1, \ldots, x_{\bar{N}}) \wedge \gamma\right\}$$
$$= g_0^\gamma(y_k) \prod_{n=1}^{\bar{N}} g_n^\gamma(y_k, x_n) \qquad (4.24)$$

for some functions g_n^γ, $n = 0, \ldots, \bar{N}$, where g_0^γ can depend on the scan data but not $\mathbf{s}_k$. For $n > 0$, $g_n^\gamma(y_k, \cdot)$ is typically the likelihood function for the measurement in y_k that is associated to target n by γ. Conditional independence means that conditioned on system state these likelihood functions are independent and combine by multiplication as shown in (4.24).

From (4.24) we can see that conditional independence means that the probability of the scan event $\{Y_k = y_k\}$, conditioned on $\mathbf{X}(t_k) = (x_1, \ldots, x_{\bar{N}})$ and γ, factors into a product of functions that each depend on the state of only one target. This type of factorization occurs when the probability distribution of the measurement due to a target is independent of all other targets. Section 4.3.2 presents an example of a scan association likelihood function that is conditionally independent.

Conditional independence implies that the likelihood function for the scan $Y_k = y_k$ in (4.11) is given by

$$L_k\left(y_k \mid \mathbf{s}_k = (x_1,\ldots,x_{\bar{N}})\right)$$
$$= \sum_{\gamma \in \Gamma_k} g_0^{\gamma}(y_k) \prod_{n=1}^{\bar{N}} g_n^{\gamma}(y_k, x_n) \Pr\{\gamma\} \quad \text{for all } \mathbf{s}_k \in \mathbf{S}^+ \tag{4.25}$$

The assumption of conditional independence of the scan association likelihood function is implicit in most multiple target trackers.

4.3.2 Scan Association Likelihood Function Example

As an example of a scan association likelihood function that is conditionally independent, we consider the following multiple target tracking situation. Detections, measurements, and false measurements are generated according to the following model.

Detections and Measurements:

Let

$$P_d\left(n \mid x\right) = \Pr\left\{\text{detecting target } n \mid \text{target } n \text{ in state } x\right\}$$
$$f\left(\theta \mid n, x\right) = \Pr\left\{\text{tgt } n \text{ produces measurement } \theta \mid \text{tgt } n \text{ detected and in state } x\right\}$$
$$\theta_k(m) = \text{ the } m\text{th measurement in scan } y_k = \left(\theta_k(1),\ldots,\theta_k(M_k)\right)$$

A measurement may be a multidimensional, nonlinear function of target state with non-Gaussian measurement error. We assume that the distribution of $\theta_k(m)$ is independent of all other measurements in the scan. The function f can depend on the sensor making the measurement although we do not indicate that dependence explicitly. We assume that $P_d(n \mid \phi) = 0$, that is, the probability of detecting a target not present is 0.

False Measurements

We assume that the number Φ of false measurements generated in a scan has a known probability distribution. False measurements, those due to noise or clutter, have probability density

$$w(z) = \mathbf{Pr}\left\{\theta_k(m) = z \mid \theta_k(m) \text{ is a false measurement}\right\} \quad \text{for } m = 1,\ldots,M_k$$

Scan Association Likelihood Function

Let $y_k = \left(\theta_k(1),\ldots,\theta_k(M_k)\right)$ be a scan with M_k contacts and γ be a scan association hypothesis. Define

$\tau(\gamma) = \{n : \gamma(m) = n > 0\}$, the set of targets associated to some contact
$\delta(\gamma) =$ number of targets in $\tau(\gamma)$ (i.e., the number of targets detected)
$\gamma^{-1}(n) =$ index of the measurement associated to target n for $n \in \tau(\gamma)$
$\varphi(\gamma) = M_k - \delta(\gamma)$, number of false measurements

Note that $\{1, ..., \bar{N}\} \setminus \tau(\gamma)$ is the set of targets not associated by γ to any contact in the scan y_k. Then

$$\ell_k \left(y_k \mid \mathbf{s}_k = (x_1, ..., x_{\bar{N}}) \wedge \gamma \right) =$$
$$\mathbf{Pr}\{\Phi = \varphi(\gamma)\} \varphi(\gamma)! \prod_{\{m : \gamma(m) = 0\}} w(\theta_k(m)) \tag{4.26}$$
$$\times \prod_{n \in \tau(\gamma)} P_d(n \mid x_n) f\left(\theta_k(\gamma^{-1}(n)) \mid n, x_n\right) \prod_{n \in \{1, ..., \bar{N}\} \setminus \tau(\gamma)} (1 - P_d(n \mid x_n))$$

To see that (4.26) holds, we break the right-hand side into three factors. The first factor,

$$\mathrm{Pr}\{\Phi = \varphi(\gamma)\} \varphi(\gamma)! \prod_{\{m : \gamma(m) = 0\}} w(\theta_k(m))$$

is the probability density of obtaining $\varphi(\gamma)$ false measurements with values $\theta_k(m)$ for $\{m : \gamma(m) = 0\}$. The factor $\varphi(\gamma)!$ is required because we are computing the probability of receiving the set of $\varphi(\gamma)$ false measurements and are not concerned with their order. (See Section 2.4 of [16].) The second factor,

$$\prod_{n \in \tau(\gamma)} P_d(n \mid x_n) f\left(\theta_k(\gamma^{-1}(n)) \mid n, x_n\right)$$

is the probability density that the targets in the set $\tau(\gamma)$ produce the measurements associated to them, and the third factor,

$$\prod_{n \in \{1, ..., \bar{N}\} \setminus \tau(\gamma)} (1 - P_d(n \mid x_n))$$

is the probability that the targets with no contacts associated to them produce no detections.

To see that (4.26) is in the form given by (4.24) and that this scan likelihood function is conditionally independent, we set

$$g_0^\gamma(y_k) = \mathrm{Pr}\{\Phi = \varphi(\gamma)\} \varphi(\gamma)! \prod_{\{m : \gamma(m) = 0\}} w(\theta_k(m))$$
$$g_n^\gamma(y_k) = P_d(n \mid x_n) f\left(\theta_k(\gamma^{-1}(n)) \mid n, x_n\right) \text{ for } n \in \tau(\gamma) \tag{4.27}$$
$$g_n^\gamma(y_k) = 1 - P_d(n \mid x_n) \text{ for } n \in \{1, ..., \bar{N}\} \setminus \tau(\gamma)$$

4.3.3 Independence Theorem

Under the assumptions of conditional independence of the scan association likelihood functions and independence of the target motion models, the independence theorem given below shows that conditioning on a data association hypothesis allows us to decompose the multiple target tracking problem into $\bar{N}$ independent single target problems. In this case, conditioning on a hypothesis greatly simplifies the joint tracking problem. In particular, we do not have to employ a joint state space representation of the target distributions.

Let $\mathbf{s}_k = (s_{1,k},...,s_{\bar{N},k})$. In addition, let $\{X_n(t); t \geq 0\}$ denote the target motion process and $q_k(s_{n,k} \mid s_{n,k-1}, n)$ the transition function at time t_k for target n for $n = 1,..., \bar{N}$.

Independence Theorem. *Suppose that the scan association likelihood functions are conditionally independent and that the prior target motion processes are mutually independent so that the multiple target transition function factors as follows:*

$$q_k\left(\mathbf{s}_k \mid \mathbf{s}_{k-1}\right) = \prod_{n=1}^{\bar{N}} q_k\left(s_{n,k} \mid s_{n,k-1}, n\right)$$

Then the posterior system state distribution conditioned on a data association hypothesis is the product of independent distributions on the targets' states.

Proof. Let $\mathbf{Y}(K) = \mathbf{y}_K = (y_1,...,y_K)$ be the scan measurements that are received at times $0 \leq t_1 \leq ... \leq t_K \leq t$. Recall that $H(k)$ is the set of all data association hypotheses on the first k scans. We wish to show for $k = 1,..., K$ that

$$p(t_k, \mathbf{s}_k \mid h) = \prod_n p_n(t_k, x_n \mid h) \text{ for } h \in H(k) \text{ and } \mathbf{s}_k = (x_1,...,x_{\bar{N}}) \in \mathbf{S}^+ \qquad (4.28)$$

where

$$p_n\left(t_k, x_n \mid h\right) = \Pr\left\{X_n(t_k) = x_n \mid h \wedge \mathbf{Y}(k)\right\} \text{ for } x_n \in S^+ \text{ and } n = 1,..., \bar{N}$$

We will prove the theorem by induction.

k=1. We first show that (4.28) holds for $k = 1$. By the independence of the prior target motion processes,

$$p(0, \mathbf{s}) = \prod_{n=1}^{\bar{N}} p_n(0, x_n) \text{ for } \mathbf{s} = (x_1,...,x_{\bar{N}}) \in \mathbf{S}^+$$

where $p_n(0,\cdot)$ is the initial state distribution on target n. Since the motion models for the targets are independent, the joint distribution at time t_1 before updating for the scan observation $Y_1 = y_1$ is

$$p^-(t_1,\mathbf{s}_1) = \prod_{n=1}^{\bar{N}} p_n^-(t_1,x_n) \text{ for } \mathbf{s}_1 = (x_1,\ldots,x_{\bar{N}}) \in \mathbf{S}^+$$

where $p_n^-(t_1,\cdot)$ is the motion-updated distribution for target n at time t_1. A data association hypothesis, $h \in H(1)$, is equal to a scan association hypothesis $\gamma \in \Gamma_1$. By the conditional independence assumption, the likelihood function for the scan Y_1 factors into functions that depend only on the state of one target and are independent of the state of the other targets.

To compute the posterior given $\mathbf{Y}(1) = Y_1$ and the association $h = \gamma$, we follow the recursion in (4.6)–(4.9) which clearly holds when we condition on a data-association hypothesis. We multiply the motion updated multiple target distribution at time t_1 by the likelihood function for $Y_1 = y_1$, both conditioned on γ, to obtain

$$\begin{aligned}
C\,p(t_1,\mathbf{s}_1 \mid \gamma) &= g_0^\gamma(y_1)\prod_{n=1}^{\bar{N}} g_n^\gamma(y_1,x_n)\prod_{n=1}^{\bar{N}} p_n^-(t_1,x_n) \\
&= g_0^\gamma(y_1)\prod_{n=1}^{\bar{N}}\left[g_n^\gamma(y_1,x_n)p_n^-(t_1,x_n)\right] \qquad (4.29) \\
&\hspace{3cm} \text{for } \mathbf{s}_1 = (x_1,\ldots,x_{\bar{N}}) \in \mathbf{S}^+
\end{aligned}$$

We obtain $p_n(t_1,x_n \mid \gamma)$, the marginal distribution on the state of target n, by integrating the right-hand side of (4.29) over all components except x_n and normalizing to obtain a probability distribution. The result is

$$p_n(t_1,x_n \mid \gamma) \propto g_n^\gamma(y_1,x_n)p_n^-(t_1,x_n) \text{ for } n = 1,\ldots,\bar{N}$$

and we see that (4.28) holds for $k = 1$.

k **implies** $k+1$. Suppose that (4.28) holds for the first k scans. Consider a hypothesis $h_{k+1} \in H(k+1)$. Then $h_{k+1} = \{h_k \wedge \gamma\}$ for some hypothesis $h_k \in H(k)$ and scan hypothesis $\gamma \in \Gamma_{k+1}$. Define

$$p^-(t_{k+1},\cdot \mid h_k) = \text{the distribution on } \mathbf{X}(t_{k+1}) \text{ given } h_k \text{ and } \mathbf{Y}(k)$$

This distribution is obtained by performing the motion and information updates for the first k scans and the motion update only from time t_k to t_{k+1}. For target n, we define

$$p_n^-(t_{k+1},\cdot \mid h_k) = \text{the distribution on } X_n(t_{k+1}) \text{ given } h_k \text{ and } \mathbf{Y}(k)$$

By assumption the target motion processes are independent. From this and the fact that (4.28) holds for k, we have

$$p^-(t_{k+1}, \mathbf{s}_{k+1} \mid h_k) = \prod_n p_n^-(t_{k+1}, x_n \mid h_k) \quad \text{for } \mathbf{s}_{k+1} = (x_1, \ldots, x_{\bar{N}}) \in \mathbf{S}^+$$

To obtain the posterior system state distribution at time t_{k+1}, we multiply $p^-(t_{k+1}, \mathbf{s}_{k+1} \mid h_k)$ by the scan likelihood function for γ_{k+1} to obtain

$$p(t_{k+1}, \mathbf{s}_{k+1} \mid h_{k+1})$$

$$= \frac{1}{C} g_0^{\gamma_{k+1}}(y_{k+1}) \prod_n g_n^{\gamma_{k+1}}(y_{k+1}, x_n) \prod_n p_n^-(t_{k+1}, x_n \mid h_k)$$

$$= \frac{1}{C} g_0^{\gamma_{k+1}}(y_{k+1}) \prod_n g_n^{\gamma_{k+1}}(y_{k+1}, x_n) p_n^-(t_{k+1}, x_n \mid h_k) \quad \text{for } \mathbf{s}_{k+1} = (x_1, \ldots, x_{\bar{N}}) \in \mathbf{S}^+$$

This shows that if (4.28) holds for k, then it is true for $k+1$. Since we have shown that (4.28) holds for $k=1$, the theorem is proved by mathematical induction.

Independent MHT Decomposition

From (4.17) and (4.28) we compute the full Bayesian posterior on the joint state space as follows.

$$p(t_K, \mathbf{s}_K) = \sum_{h \in H(K)} \alpha(h \mid \mathbf{y}_K) p(t_K, \mathbf{s}_K \mid h)$$

$$= \sum_{h \in H(K)} \alpha(h \mid \mathbf{y}_K) \prod_n p_n(t_K, x_n \mid h) \quad \text{for } \mathbf{s}_K = (x_1, \ldots, x_{\bar{N}}) \in \mathbf{S}^+ \tag{4.30}$$

This is the independent MHT decomposition.

Marginal posteriors can be computed in a similar fashion. Let $\bar{p}_n(t_K, \cdot)$ be the marginal posterior on $X_n(t_K)$ for $n = 1, \ldots, \bar{N}$. Then

$$\bar{p}_n(t_K, x_n) = \int \left[\sum_{h \in H(K)} \alpha(h \mid \mathbf{y}_K) \prod_{l=1}^{\bar{N}} p_l(t_K, x_l \mid h) \right] \prod_{l \neq n} dx_l$$

$$= \sum_{h \in H(K)} \alpha(h \mid \mathbf{y}_K) p_n(t_K, x_n \mid h) \int \prod_{l \neq n} p_l(t_K, x_l \mid h) dx_l$$

$$= \sum_{h \in H(K)} \alpha(h \mid \mathbf{y}_K) p_n(t_K, x_n \mid h) \quad \text{for } n = 1, \ldots, \bar{N}$$

Thus the posterior marginal distribution on target n may be computed as the weighted sum of the posterior distributions for target n conditioned on h.

4.3.4 Independent MHT Recursion

Let $q_0(x \mid n) = \mathbf{Pr}\{X_n(0) = x\}$ and $q_k(x \mid x', n) = \mathbf{Pr}\{X_n(t_k) = x \mid X_n(t_{k-1}) = x'\}$. Under the assumptions of the independence theorem, the independent MHT recursion given below holds.

Independent MHT Recursion

Let $\mathbf{y}_k = (y_1, \ldots, y_k)$ be the measurements in the first k scans and $H(0) = \{h_0\}$ where h_0 is the empty set (i.e., the hypothesis with no associations). Set $t_0 = 0$, $\alpha(h_0) = 1$, and

$$p_n(t_0, x \mid h_0) = q_0(x \mid n) \text{ for } x \in S^+ \text{ and } n = 1, \ldots, \bar{N}$$

Compute conditional target distributions. For $k = 1, 2, \ldots$, do the following: For each $h_k \in H(k)$, find the $h_{k-1} \in H(k-1)$ and $\gamma \in \Gamma_k$ such that $h_k = h_{k-1} \wedge \gamma$. Then compute

$$p_n^-(t_k, x \mid h_{k-1}) = \int_{S^+} q_k(x \mid x', n) \, p_n(t_{k-1}, x' \mid h_{k-1}) \, dx' \text{ for } n = 1, \ldots, \bar{N}$$

$$p_n(t_k, x \mid h_k) = \frac{1}{c(n, h_k)} g_n^\gamma(y_k, x) p_n^-(t_k, x \mid h_{k-1}) \text{ for } x \in S^+ \text{ and } n = 1, \ldots, \bar{N} \quad (4.31)$$

$$p(t_k, \mathbf{s}_k \mid h_k) = \prod_n p_n(t_k, x_n \mid h_k) \text{ for } \mathbf{s}_k = (x_1, \ldots, x_{\bar{N}}) \in \mathbf{S}^+$$

where $c(n, h_k)$ is the constant that makes $p_n(t_k, \cdot \mid h_k)$ a probability distribution.

Compute association probabilities. For $k = 1, 2, \ldots$ and $h_k = h_{k-1} \wedge \gamma \in H(k)$ compute

$$\beta(h_k) = \alpha(h_{k-1} \mid \mathbf{y}_{k-1}) \mathbf{Pr}\{\gamma\} \int g_0^\gamma(y_k) \prod_n g_n^\gamma(y_k, x_n) p_n^-(t_k, x_n \mid h_{k-1}) dx_1 \cdots dx_{\bar{N}}$$

$$= \alpha(h_{k-1} \mid \mathbf{y}_{k-1}) \mathbf{Pr}\{\gamma\} g_0^\gamma(y_k) \prod_n \int g_n^\gamma(y_k, x_n) p_n^-(t_k, x_n \mid h_{k-1}) dx_n \quad (4.32)$$

Then

$$\alpha(h_k \mid \mathbf{y}_k) = \frac{\beta(h_k)}{\displaystyle\sum_{h_k' \in H(k)} \beta(h_k')} \text{ for } h_k \in H(k) \quad (4.33)$$

In (4.31), the independent MHT recursion performs a motion update of the probability distribution on target n given h_{k-1} and multiplies the result by $g_n^\gamma(y_k, x)$, the likelihood function of the measurement associated to target n by γ. When this product is normalized to a probability distribution, we obtain the posterior on target n given $h_k = h_{k-1} \wedge \gamma$. Note that this computation is performed independently of the other targets. This is where the independent MHT obtains its power and simplicity. Only the computation of the association probabilities in (4.32) and (4.33) requires interaction with the other targets and the likelihoods of the measurements associated to them.

To see that (4.33) holds, we observe that by (4.22) and (4.23)

$$\alpha(h_{k-1} \mid y_{k-1}) = c_1 C(h_{k-1}) \text{ and } \beta(h_k) = c_2 C(h_k)$$

where c_1 does not depend on h_{k-1} and c_2 does not depend on h_k. Thus (4.33) follows from (4.23).

Usually, one does not complete the computation of the full joint posterior distribution given by (4.17). Often, only the (independent) single target distributions resulting from the hypothesis with the highest probability are displayed.

The recursion does not say how to construct the data association hypotheses $h_k = h_{k-1} \wedge \gamma$. An algorithm that implements the recursion will have to specify a way to do this. We give an example of how to do this in Section 4.4 where we describe an MHT algorithm for the linear-Gaussian case.

4.4 LINEAR-GAUSSIAN MULTIPLE HYPOTHESIS TRACKING

In this section we consider the special case that is used most often in MHT implementations. This is the independent MHT case in which the prior target processes $\{X_n(t); t \geq 0\}$ for $n = 1, \ldots, \overline{N}$ have l-dimensional Gaussian distributions at each time $t \geq 0$. The observations are linear functions of the target state with Gaussian errors. At time t, an hypothesis h associates each contact in the scan with a target or false measurement. If a contact is associated to a target, we perform the state estimation for the target using a standard Kalman filter. The result is a set of Gaussian densities representing the state estimates for each target. If a target has no contacts assigned to it under data association hypothesis h, then it is treated as "not tracked," which means it has the prior distribution given in (4.34). Under hypothesis h, the joint target state distribution is the product of these independent Gaussian distributions. Usually, these independent target distributions are displayed by showing the 2-σ ellipses corresponding to the marginal position distribution for each target. Typically, these are overlaid on a geographic display.

4.4.1 MHT Recursion for Linear-Gaussian Case

We now describe a recursive procedure for implementing an independent MHT in the linear-Gaussian case when all targets have identical motion models and detection characteristics.

For this procedure, we suppose that the assumptions of the scan association likelihood example in Section 4.3.2 hold and in particular that the scan association likelihood function is given by (4.26). Let $\mathcal{N}(\mu_0, P_0)$ be the prior distribution on S for each target given it is present. Let $p(\phi)$ be the prior probability that target n is not present. The prior distribution for target n at time $t = 0$ is

$$
p_n(0, x) = \begin{cases} (1 - p(\phi))\eta(x, \mu_0, P_0) & \text{for } x \in S \\ p(\phi) & \text{for } x = \phi \end{cases} \tag{4.34}
$$

Since the prior distributions on the targets are identical, they are indistinguishable to the tracker. If h and h' are association hypotheses such that h' can be obtained from h by permuting the target number indices, then $\alpha(h) = \alpha(h')$. In fact, the joint posterior distribution will be symmetric under permutation of the target indices. This simply reflects the fact that the numbering of the targets is arbitrary when they are all identical. With this in mind, we may specify an arbitrary target numbering and compute the posterior based on that numbering. The recursion described below contains a method for specifying this "arbitrary numbering." When the priors on the targets are distinct for each target, this problem does not occur.

For the kth scan, the contacts are numbered from 1 to M_k, the number of contacts in scan k. For the first scan, an association, γ, designates those contacts that are real and those that are false measurements. Let $\varphi(\gamma)$ be the number of contacts designated as false. Once this designation is made, the association is completed by assigning the remaining contacts to targets $n = 1, \ldots, M_1 - \varphi(\gamma)$ in order of their contact number. The lowest contact number is assigned to target 1, and so on. (Recall that in a scan, no target generates more than one contact and no contact is due to more than one target. There are no split or merged measurements.) The Bayesian information update is performed for targets 1 through $M_1 - \varphi(\gamma)$, and the target distributions are motion updated to the time of the next scan. The targets that have been assigned contacts in an hypothesis h are called *existing targets in h.*

Now consider what happens when scan 2 is received. Let $h_1 \in H(1)$. To extend h_1 to an hypothesis $h_2 = h_1 \wedge \gamma$ where γ is a scan association hypothesis for scan 2, we will allow only certain scan association hypotheses. In particular, γ must have the property that contacts that are designated as real but that are not assigned by γ to an existing target in h_1, are assigned to new targets in order of their contact number. Thus, the lowest contact number not assigned to an existing target is assigned to the lowest new target number. Each hypothesis in $H(1)$ is extended to the set of hypotheses in $H(2)$ that can be obtained by "adding" all allowable scan associations for that hypothesis. This forms the set $H(2)$ of hypotheses on the first two scans. This same process is repeated for each scan to obtain $H(3)$, $H(4)$, and so on.

4.4.2 Posterior Distributions and Association Probabilities

We now show how the independent MHT recursion is performed for this linear-Gaussian case. Assume that the conditions of the independence theorem are satisfied. Then, conditioned on an hypothesis h, the posterior distribution on the joint target state space is the product of independent distributions on each target. Thus at time t_{k-1}, the posterior distribution is given by

$$
\begin{aligned}
p(t_{k-1}, \mathbf{s}_{k-1}) &= \sum_{h \in H(k-1)} \alpha(h \mid \mathbf{y}_{k-1}) p(t_{k-1}, \mathbf{s}_{k-1} \mid h) \\
&= \sum_{h \in H(k-1)} \alpha(h \mid \mathbf{y}_{k-1}) \prod_n p_n(t_{k-1}, x_n \mid h) \ \text{ for } \mathbf{s}_{k-1} = (x_1, \ldots, x_{\bar{N}}) \in \mathbf{S}^+
\end{aligned}
$$

Updating for target motion to time t_k, we have

$$
p_n^-(t_k, x \mid h) = \begin{cases} \eta\left(x, \mu_k^-(n), P_k^-(n)\right) & \text{if } n \text{ is an existing target in } h \\ p_n(0, \cdot) & \text{if no contacts have been associated to target } n \text{ in } h \end{cases} \tag{4.35}
$$

where $\mu_k^-(n)$ and $P_k^-(n)$ are the mean and covariance of the motion-updated distribution for target n obtained from the Kalman filter recursion in Chapter 3. If a target has no contacts associated to it by h, we take its posterior distribution to be equal to the prior at time 0. In particular, we do not apply the motion model to a target that has no contacts associated to it. Assume that we have calculated the association probabilities $\alpha(h)$ for $h \in H(k-1)$.

Let $y_k = (\theta_k(1), \ldots, \theta_k(M_k))$ be the scan of M_k measurements at time t_k. We consider the measurement $\theta_k(m)$ to be a realization of the random variable $\Theta_k(m)$. If $\Theta_k(m)$ is generated by target n, then

$$
\Theta_k(m) = H_m X_n(t_k) + \varepsilon_m
$$

where

$$
X_n(t_k) \sim \mathcal{N}(\mu_n, P_n), \ \varepsilon_m \sim \mathcal{N}(0, R_m)
$$

and H_m is a matrix, and ε_m is independent of X_k. The matrices H_m and R_m can depend on k, but we do not indicate this dependence. Assume that the detection probability $P_d(n \mid x) = p_d$ is independent of n and x and that the number of false measurements, Φ, detected in the scan follows the Poisson distribution,

$$\mathbf{Pr}\{\Phi = j\} = \frac{(\rho V)^j}{j!} e^{-\rho V} \text{ for } j = 0, 1 \ldots$$

where ρ is the false measurement density. The probability density for obtaining the measurement $\theta_k(m) = z$ given that it results from a false measurement is $w(z) = 1/V$ where V is the volume of the measurement space.

Let $h_k = h_{k-1} \wedge \gamma$ where $h_{k-1} \in H(k-1)$ and $\gamma \in \Gamma_k$ is an allowable scan association hypothesis for h_{k-1}. Note that γ may assign one or more contacts to new targets, ones that do not exist in h_{k-1}. Let N_k be the number of existing targets in h_k. From (4.26) we obtain

$$\ell_k\left(y_k \mid \gamma \wedge \mathbf{s}_k = (x_1, \ldots, x_{\bar{N}})\right)$$
$$= \rho^{\varphi(\gamma)} e^{-\rho V} \left(1 - p_d\right)^{N_k - \delta(\gamma)} p_d^{\delta(\gamma)} \prod_{n \in \tau(\gamma)} f\left(\theta_k\left(\gamma^{-1}(n)\right) \mid n, x\right) \qquad (4.36)$$

where

$\tau(\gamma) = \{n : \gamma(m) = n > 0\}$, the set of targets associated to some contact
$\delta(\gamma) = $ the number of targets in $\tau(\gamma)$ (i.e., the number of targets detected)
$\gamma^{-1}(n) = $ index of the measurement associated to target n for $n \in \tau(\gamma)$
$\varphi(\gamma) = M_k - \delta(\gamma)$, the number of false measurements

and

$$f(y \mid n, x) = \eta(y, H_{m_n} x, R_{m_n}) \text{ where } m_n = \gamma^{-1}(n)$$

In order to use the independent MHT recursion, we put ℓ_k into the form given in (4.24). Letting

$$g_0^\gamma(y_k) = \rho^{\varphi(\gamma)} e^{-\rho V} p_d^{\delta(\gamma)} \left(1 - p_d\right)^{N_k - \delta(\gamma)}$$
$$g_n^\gamma(y_k, x) = \begin{cases} \eta(y, H_{\gamma^{-1}(n)} x, R_{\gamma^{-1}(n)}) & \text{if } n \in \tau(\gamma) \\ 1 & \text{otherwise} \end{cases} \qquad (4.37)$$

we have

$$\ell_k(y_k \mid \mathbf{s}_k = (x_1, \ldots, x_{\bar{N}}) \wedge \gamma) = g_0^\gamma(y_k) \prod_{n=1}^{N_k} g_n^\gamma(y_k, x_n)$$

4.4.2.1 Posterior Target Distributions

From (4.31), we have that

$$p_n(t_k, x \mid h_k) \propto g_n^{\gamma}(y_k, x) p_n^-(t_k, x \mid h_{k-1})$$
$$\propto f\left(\theta_k\left(\gamma^{-1}(n)\right) \mid n, x\right) p_n^-(t_k, x \mid h_{k-1}) \text{ for } n \in \tau(\gamma) \qquad (4.38)$$

Since

$$p_n^-(t_k, x \mid h_{k-1}) = \eta\left(x, \mu_k^-(n), P_k^-(n)\right)$$

the right-hand side of (4.38) is the joint density function, $f(x, y)$, of $X_n(t_k)$ and $\Phi_k(m)$ evaluated at $(x, \theta_k(m))$. Since the value of y is fixed at $\theta_k(m)$, the right-hand side of (4.38) is proportional to the conditional density of $X_n(t_k)$ given $\Theta_k(m) = \theta_k(m)$. By (A.5) of the Gaussian density lemma,

$$p_n(t_k, x \mid h_k) = \eta\left(x, \mu_k(n), P_k(n)\right)$$

where

$$m_n = \gamma^{-1}(n)$$

$$\mu_k(n) = \mu_k^-(n) + P_k^-(n) H_{m_n}^T \left(H_{m_n} P_k^-(n) H_{m_n}^T + R_{m_n}\right)^{-1} \left(\theta_k(m_n) - H_{m_n} \mu_k^-(n)\right)$$

$$P_k(n) = P_k^-(n) - P_k^-(n) H_{m_n}^T \left(H_{m_n} P_k^-(n) H_{m_n}^T + R_{m_n}\right)^{-1} H_{m_n} P_k^-(n)$$

The result of the above procedure is

$$p_n(t_k, x \mid h_k) = \begin{cases} \eta(x, \mu_k(n), P_k(n)) \text{ for } n \in \tau(\gamma) \\ p_n^-(t_k, x) \qquad \text{for } n \in \{1, \ldots \bar{N}\} \setminus \tau(\gamma) \end{cases} \qquad (4.39)$$

4.4.2.2 Association Probabilities

From (4.32) and (4.37) we have

$$\beta(h_k) = \alpha(h_{k-1} \mid \mathbf{y}_{k-1}) \Pr\{\gamma\} \rho^{\varphi(\gamma)} e^{-\rho V} \left(1 - P_d\right)^{N_k - \delta(\gamma)} P_d^{\delta(\gamma)}$$
$$\times \prod_{n \in \tau(\gamma)} \int f\left(\theta_k\left(\gamma^{-1}(n)\right) \mid n, x\right) p_n^-(t_k, x \mid h_{k-1}) \, dx \qquad (4.40)$$

Setting

$$f(y \mid x) = f(y \mid n, x) \text{ and } f_1(x) = p_n^-(t_k, x \mid h_{k-1})$$

in the Gaussian density lemma and $m_n = \gamma^{-1}(n)$, we have by (A.2) and (A.4) that

$$\int f(\theta_k(m_n) \mid n, x) \, p_n^-(t_k, x \mid h_{k-1}) dx$$

$$= \int f(x, \theta_k(m_n)) dx \qquad (4.41)$$

$$= f_2(\theta_k(m_n))$$

$$= \eta\left(\theta_k(m_n), H_{m_n} \mu_n, H_{m_n} P_k^-(n) H_{m_n}^{\ T} + R_{m_n}\right)$$

Thus

$$\beta(h_k) = \alpha(h_{k-1} \mid y_{k-1}) \Pr\{\gamma\} \rho^{\varphi(\gamma)} e^{-\rho V} \left(1 - p_d\right)^{N_k - \delta(\gamma)} p_d^{\delta(\gamma)}$$

$$\times \prod_{n \in \tau(\gamma)} \eta\left(\theta_k(m_n), H_{m_n} \mu_n, H_{m_n} P_k^-(n) H_{m_n}^{\ T} + R_{m_n}\right)$$

Now we can compute

$$\alpha(h_k \mid y_k) = \frac{\beta(h_k)}{\displaystyle\sum_{h \in H(k)} \beta(h)} \quad \text{for } h_k \in H(k) \qquad (4.42)$$

4.5 NONLINEAR JOINT PROBABILISTIC DATA ASSOCIATION

This section presents an extension of the JPDA approximation to multiple target tracking given in [2, 12]. For readers familiar with JPDA, nonlinear JPDA can be thought of as an extension of JPDA that relaxes the linear-Gaussian assumptions to allow for nonlinear measurements with non-Gaussian errors as well as non-Gaussian target state distributions. This extension is implemented using particle filters. We extend JPDA in one other respect. We allow for the creation of new target tracks and the deletion of existing ones.

In JPDA we make the approximation that $p^-(t_k, s_k)$, the motion-updated joint target state distribution at time t_k, contains all the information that we need from the past to compute the posterior at time t_k. In particular, we do not need to reconsider past scans and association hypotheses. We further assume that the scan association likelihood functions are conditionally independent and that the posterior target state distributions are mutually independent.

4.5.1 Scan Association Hypotheses

As above we assume that the measurements at time t_k arrive in a scan of contacts. In contrast to MHT, we make the approximation that the joint distribution on target state, obtained from the scans at times $t_1, \ldots, t_{k-1}$ and motion-updated to time t_k, is a sufficient statistic for computing the association probabilities for the scan at time t_k and the posterior joint target state distribution at time t_k.

In MHT we form data association hypotheses by assigning contacts to specific targets or to false measurements. Within a data association hypothesis, a contact is associated to a target with probability 1 or 0. This is a *hard* association. For JPDA we relax this assumption and allow *soft* associations, that is, ones where the probability of a contact being associated to a target is equal to p where $0 \leq p \leq 1$. When we update the target distribution for the effect of this contact, we approximate the posterior target state distribution as the weighted mixture of the distribution updated with the contact information and the distribution not updated with the contact information using the weights p and $1 - p$.

For a single scan with even a modest number of contacts, there will be a large number of association hypotheses. JPDA computes the scan association probabilities for each of these hypotheses and then computes a_{mn}, the probability measurement m is associated to target n, by summing these probabilities over all scan associations in which measurement m is associated to target n. It replaces the set of association hypothesis with a single soft-association hypothesis with association probabilities given by the matrix (a_{mn}). From this it computes the joint target state distribution as a mixture of distributions obtained from hard associations.

Let $y_k = (\theta_k(1), \ldots, \theta_k(M_k))$ be the scan of contacts at time t_k. Recall that a scan association γ is function defined on $\{1, \ldots, M_k\}$ such that

$$\gamma(m) = \begin{cases} n > 0 & \text{if contact } m \text{ is associated to target } n \\ 0 & \text{if contact } m \text{ is associated to a false measurement} \end{cases}$$

and each target has at most one contact associated to it. As before, Γ_k is the set of scan associations for the scan y_k.

4.5.2 Scan Association Probability

Let

$$\alpha(\gamma \mid y_k) = \Pr\{\gamma \mid Y_k = y_k\} \text{ for } \gamma \in \Gamma_k$$

Since $p^-(t_k, \mathbf{s}_k)$ a sufficient statistic for computing the posterior at time t_k, it follows that in (4.22), $C(h_{k-1}) = c$ where c does not depend on h_{k-1}, and that $p^-(t_k, \mathbf{s}_k \mid h_{k-1})$ does not depend on h_{k-1}. As a result (4.22) becomes

$$C(\gamma) = c \Pr\{\gamma\} \int \ell_k(y_k \mid \gamma \wedge \mathbf{s}_k) p^-(t_k, \mathbf{s}_k) d\mathbf{s}_k \tag{4.43}$$

where $\ell_k(y_k \mid \gamma \wedge \mathbf{s}_k) = \Pr\{Y_k = y_k \mid \gamma \wedge \mathbf{X}(t_k) = \mathbf{s}_k\}$ is the scan association likelihood function for γ. Define

$$\hat{C}(\gamma) = \Pr\{\gamma\} \int \ell_k(y_k \mid \gamma \wedge \mathbf{s}_k) p^-(t_k, \mathbf{s}_k) d\mathbf{s}_k \tag{4.44}$$

From (4.23) we have

$$\alpha(\gamma \mid y_k) = \frac{\hat{C}(\gamma)}{\displaystyle\sum_{\gamma' \in \Gamma_k} \hat{C}(\gamma')} \tag{4.45}$$

Suppose we have N_k tracks (targets present) at time t_k and let $\mathbf{s}_k = (x_1, \ldots, x_{N_k})$. Note, we do not indicate the components for the targets not present. Each one of these components will have $\Pr\{\phi\} = 1$ and will not affect the computations below. Since the target state distributions are independent

$$p^-(t_k, \mathbf{s}_k) = \prod_{n=1}^{N_k} p_n^-(t_k, x_n)$$

Since the scan likelihood function is conditionally independent, we have

$$\ell_k(y_k \mid \gamma \wedge \mathbf{s}_k) = g_0^\gamma(y_k) \prod_{n=1}^{N_k} g_n^\gamma(y_k, x_n) p_n^-(t_k, x_n)$$

and

$$\hat{C}(\gamma) = \Pr\{\gamma\} g_0^\gamma(y_k) \prod_{n=1}^{N_k} \int g_n^\gamma(y_k, x_n) p_n^-(t_k, x_n) dx_n \tag{4.46}$$

Association Example

Suppose that the scan association likelihood function is the one given in (4.27) with the detection probability equal to a constant p_d and the false measurements modeled by a Poisson point process with a constant intensity ρ over the measurement space which has volume V. In this case

$$\Pr\{\Phi = j\} = \frac{(\rho V)^j e^{-\rho V}}{j!} \quad \text{and} \quad w(z) = V^{-1} \text{ for all } z$$

Then (4.27) becomes

$$g_0^\gamma(y_k) = e^{-\rho V} \rho^{\varphi(\gamma)}$$
$$g_n^\gamma(y_k) = p_d f\left(\theta_k\left(\gamma^{-1}(n)\right) \mid n, x_n\right) \text{ for } n \in \tau(\gamma) \tag{4.47}$$
$$g_n^\gamma(y_k) = 1 - p_d \text{ for } n \in \{1, \dots, N_k\} \backslash \tau(\gamma)$$

Recall that $\delta(\gamma)$ equals the number of detected targets in hypothesis γ so that

$$\ell_k\left(y_k \mid \gamma \wedge s_k\right) = e^{-\rho V} \rho^{\varphi(\gamma)} \left(1 - p_d\right)^{N_k - \delta(\gamma)} p_d^{\delta(\gamma)} \prod_{n \in \tau(\gamma)} f\left(\theta_k\left(\gamma^{-1}(n)\right) \mid n, x_n\right)$$

$$\hat{C}(\gamma) = \Pr\{\gamma\} e^{-\rho V} \rho^{\varphi(\gamma)} \left(1 - p_d\right)^{N_k - \delta(\gamma)} p_d^{\delta(\gamma)}$$
$$\times \prod_{n \in \tau(\gamma)} \int f\left(\theta_k\left(\gamma^{-1}(n)\right) \mid n, x_n\right) p_n^-(t_k, x_n) dx_n \tag{4.48}$$

and $\alpha(\gamma \mid y_k)$ is computed by (4.45).

Linear-Gaussian Example

Suppose that $P\{\gamma\}$ is equal to a constant independent of γ and that the linear-Gaussian assumptions of Section 4.4 hold. In particular, let $m_n = \gamma^{-1}(n)$ and suppose that

$$p_n^-(t_k, x_n) = \eta\left(x_n, \mu_n^-, P_n^-\right) \quad \text{and} \quad f(y \mid n, x) = \eta\left(y, H_{m_n} x, R_{m_n}\right)$$

then by (4.45) and (4.48)

$$\alpha(\gamma \mid y_k) = c_1 \rho^{\varphi(\gamma)} \left(1 - p_d\right)^{N_k - \delta(\gamma)} p_d^{\delta(\gamma)}$$
$$\times \prod_{n \in \tau(\gamma)} \eta\left(\theta_k(m_n), H_{m_n} \mu_n^-, H_{m_n} P_n^- H_{m_n}^T + R_{m_n}\right) \tag{4.49}$$

where c_1 is a constant that does not depend on γ. This matches the association probability given in in (6.2.5-2) of Section 6.2.5 of [12] which derives the parametric form of JPDA.

4.5.3 JPDA Posterior

In order to compute the JPDA posterior, we compute the soft-association matrix. Let

$$a_{mn} = \sum_{\{\gamma:\gamma(m)=n\}} \alpha(\gamma \mid y_k) \tag{4.50}$$

Then a_{mn} is the sum of the association probabilities $\alpha(\gamma \mid y_k)$ for which γ associates contact m to target n and (a_{mn}) is the soft-association matrix for scan y_k. Let

$$\bar{a}_n = 1 - \sum_{m=1}^{M_k} a_{mn}$$

Then $\bar{a}_n$ is the probability that no contact in scan y_k is associated to target n. Let

$$p_n^+(t_k, x \mid m) = \text{posterior density for target } n \text{ updated with measurement } m$$

Then the JPDA posterior for target n is computed by

$$p_n(t_k, x) = \bar{a} p_n^-(t_k, x) + \sum_{m=1}^{M_k} a_{mn} p_n^+(t_k, x \mid m) \tag{4.51}$$

In standard JPDA, the distribution in (4.51) is approximated by a Gaussian one.

4.5.4 Allowing New Targets and Deleting Existing Ones

In the standard JPDA the number of targets is assumed known and fixed. We extend that by allowing for the creation of new target tracks and the deletion of existing ones.

If N_k is the number of targets present at time t_k, we add an additional potential target with density function $p^*(x)$ for $x \in S$, which is typically taken as the distribution on the state of an undetected target. The appropriate distribution will depend on the specifics of the tracking problem. Its choice tends to be a matter of art rather than science. One then computes the association probabilities with this additional target $N_k + 1$ added and the probability $\bar{a}_{N_k+1}$ that no contact in the scan associates to this target. If this probability is less than a threshold value α_1, then target $N_k + 1$ is added to the list of existing targets and the posterior distribution is calculated by (4.51). If it is not greater than α_1, then recompute the association probabilities with N_k targets and compute the posterior by (4.51). As an approximation, one could simply remove column N_{k+1} from the soft-association matrix and compute the posterior.

To delete an existing target, one sets a threshold value α_0 and a number of scans, K. If $\bar{a}_n > \alpha_0$ for K consecutive scans, then delete target n. Clearly, setting the values of α_0, α_1, and K is a matter of art rather than science. Some experimentation will be required for each situation.

4.5.5 Particle Filter Implementation

In this section, we describe a particle filter implementation for nonlinear JPDA. For this discussion we assume that the reader is familiar with a particle filter implementation for a single target tracker such as the one described in Section 3.3.

We begin at time t_k and suppose that there are N_k targets whose distributions are represented by particle filters. For target n the distribution is represented by a set of J particles and weights

$$\left\{ \left(x_n^j(t_k), w_n^j(t_{k-1}) \right); j = 1, \ldots, J \right\} \tag{4.52}$$

The weight in (4.52) has time index t_{k-1} because we are starting with the distribution for target n conditioned on scans $\{y_1, \ldots, y_{k-1}\}$ and motion-updated to time t_k.

If we are at the time t_1 of the first scan, then some initiation logic is required. One could for example initiate a track from each contact in the scan. Those that are initiated from false measurements should eventually be dropped by the deletion logic. Often one does not display tracks until they have had contacts "associated" to them in three or more consecutive scans. In JPDA this criterion could be implemented by setting a threshold α_0 and deleting target n, if $\bar{a}_n > \alpha_0$ for the first three scans. The targets that are not deleted would be displayed and carried as existing targets for the JPDA computations in the next scan.

Let $\{(x_0^j, w_0^j); j = 1, \ldots, J\}$ be the particle filter representation of the distribution of the "undetected target" discussed in Section 4.5.4. Add this as target $N_k + 1$ by setting $x_j^{N_k+1}(t_k) = x_0^j$ and $w_j^{N_k+1}(t_{k-1}) = w_0^j$ for $j = 1, \ldots, J$. Then compute the JPDA posterior according to the JPDA particle filter posterior algorithm given below. By applying the tests described in Section 4.5.4, one can decide whether to create a new target or delete an existing target.

JPDA Particle Filter Algorithm

Let $y_k = (\theta_k(1), \ldots, \theta_k(M_k))$ be the scan of measurements at time t_k.

For $n = 1, \ldots, N_k + 1$ and $m = 1, \ldots, M_k$, compute

$$c^+(m,n) = \sum_{j=1}^{J} P_d\left(n \mid x_n^j(t_k)\right) f\left(\theta_k(m) \mid n, x_n^j(t_k)\right) w_n^j(t_{k-1})$$

$$c^-(n) = \sum_{j=1}^{J}\left(1 - P_d\left(n \mid x_n^j(t_k)\right)\right) w_n^j(t_{k-1}) \tag{4.53}$$

For each association hypothesis γ, compute

$$g_0^\gamma(y_k) = \Pr\{\Phi = \varphi(\gamma)\} \varphi(\gamma)! \prod_{\{m:\gamma(m)=0\}} w(\theta_k(m))$$

$$\hat{C}(\gamma) = \Pr\{\gamma\} g_0^\gamma(y_k) \prod_{n \in \tau(\gamma)} c^+\left(\gamma^{-1}(n), n\right) \prod_{n \in \{1,\ldots,M_k+1\}\backslash\tau(\gamma)} c^-(n) \tag{4.54}$$

$$\alpha(\gamma \mid y_k) = \frac{\hat{C}(\gamma)}{\sum_{\gamma' \in \Gamma_k} \hat{C}(\gamma')}$$

For each measurement m and target n compute

$$a_{mn} = \sum_{\{\gamma:\gamma(m)=n\}} \alpha(\gamma \mid y_k) \text{ and } \bar{a}_n = 1 - \sum_{m=1}^{M_k+1} a_{mn}$$

$$\left\{\omega_+^j(n \mid m); j = 1, \ldots, J\right\} = \text{ the posterior weights from updating} \atop \text{target } n \text{ with measurement } m \tag{4.55}$$

For each target n compute the posterior particle weights

$$w_n^j(t_k) = \bar{a}_n w_n^j(t_{k-1}) + \sum_{m=1}^{M_k} a_{mn}\omega_+^j(n \mid m) \text{ for } j = 1, \ldots, J \tag{4.56}$$

4.5.6 Example

Section 1.5 provides an example of using particle filters to apply nonlinear JPDA to a simple multiple target tracking problem involving two targets and no false alarms. For that example we assumed that the number of targets was fixed and known. Scans consisted of a single contact and there are only two association hypotheses for each scan,

$$\gamma_1 = \text{ contact associated to target } 1$$
$$\gamma_2 = \text{ contact associated to target } 2$$

We assume that

$$\Pr\{\gamma_1\} = \Pr\{\gamma_2\} = 1/2$$

and that the detection probability equals a constant p_d. As a result

$$c^+(1,n) = p_d \sum\nolimits_{j=1}^{J} f\big(\theta_k(1) \mid n, x_j^n(t_k)\big) w_j^n(t_{k-1})$$

$$c^-(n) = 1 - p_d$$

$$\hat{C}(\gamma_n) = c^+(1,n)(1 - p_d)$$

$$\alpha(\gamma_n) = \frac{c^+(1,n)}{c^+(1,1) + c^+(1,2)}$$

where $f = p_d L$ and L is the likelihood function defined by (1.7). Notice that the factor p_d cancels from the numerator and denominator of the expression for $\alpha(\gamma_n)$ so that $\alpha(\gamma_n) = \alpha(n)$ in Section 1.5.1.

The matrix of association probabilities (a_{mn}) is a column vector $(a_{11}, a_{12})^T$ where a_{1n} is the probability that the contact is associated to target n. Since there are no false measurements and exactly two targets $\bar{a}_n = 1 - a_{1n}$. Thus (4.56) becomes

$$w_j^n(t_k) = a_{1n}\omega_j^+(n \mid 1) + (1 - a_{1n}) w_j^n(t_{k-1})$$

in parallel with (1.9).

4.6 PROBABILISTIC MULTIPLE HYPOTHESIS TRACKING

Probabilistic multiple hypothesis tracking (PMHT), first developed in [15, 17], relaxes the MHT scan assumptions in two ways. First it allows for soft associations, and second it allows more than one contact per scan to be associated to a target. PMHT assumes that the number of targets N is known and fixed throughout the interval $[0, T]$ during which scans are received. For each measurement in a scan, it assumes that there is a probability distribution on the target generating the measurement. Specifically, if $\theta_k(m)$ is a measurement in scan θ_k and γ_k specifies the association of measurements to targets for scan θ_k, then $\Pr\{\gamma_k(m) = n\} = \pi_{kn}$ for $n = 1, \dots, N$, and the association of measurement m to a target is independent of the association of any other measurement in scan k. One can think of the association process for scan k as proceeding by making an independent draw from the above distribution to determine the target to be associated to each measurement in the scan. This process may yield more than one measurement associated to the same target, which violates the standard scan association assumptions. The benefit of this relaxation is that the computational load for PMHT grows linearly in the number of measurements and targets. This contrasts with the exponential computational growth for MHT, which necessitates pruning of the hypotheses to keep the computational load manageable.

There is another crucial difference that distinguishes PMHT from MHT and JPDA. PMHT computes the maximum a posteriori probability (MAP) path for each of the N targets but does not provide the target state distributions.

In the remainder of this section we describe the assumptions made by PMHT and describe a nonlinear method of computing the PMHT solution that may be implemented by a particle filter. Then we present the special case of the PMHT algorithm that holds under linear-Gaussian assumptions.

4.6.1 PMHT Assumptions

PMHT makes the following assumptions.

4.6.1.1 Target Motion Model

There are N targets and this number is fixed and known. The stochastic process $\{\mathbf{X}(t); t \geq 0\}$ representing the motion of these targets is composed of N independent Markov processes $\{X_n(t); t \geq 0\}$ for $n = 1, \ldots, N$ so that

$$\mathbf{X}(t) = \left(X_1(t), \ldots, X_N(t) \right) \text{ for } t \geq 0 \text{ and } n = 1, \ldots, N$$

For $n = 1, \ldots, N$ and $k = 2, \ldots, K$, let

$$q\left(x_n(1) \right) = \Pr\left\{ X_n(t_1) = x_n(1) \right\}$$
$$q_k\left(x_n(k) \mid x_n(k-1) \right) = \Pr\left\{ X_n(t_k) = x_n(k) \mid X_n(t_{k-1}) = x_n(k-1) \right\}$$

Then

$$\Pr\left\{ \mathbf{X}(t_k) = \left(x_1(k), \ldots, x_N(k) \right) \text{ for } k = 1, \ldots, K \right\}$$
$$= \prod_{n=1}^{N} \left[q\left(x_n(1) \right) \prod_{k=2}^{K} q_k\left(x_n(k) \mid x_n(k-1) \right) \right] \tag{4.57}$$

Define

$$x_n = \left(x_n(1), \ldots, x_n(K) \right), \ \mathbf{x} = \left(x_1, \ldots, x_N \right), \text{ and } \mathbf{x}(k) = \left(x_1(k), \ldots, x_N(k) \right) \tag{4.58}$$

Then x_n specifies the path of target n, and $\mathbf{x}$ specifies the paths of all N targets over time while $\mathbf{x}(k)$ gives the states of the N targets at time t_k.

4.6.1.2 Measurements

At each time t_k there is a scan

$$\theta_k = \left(\theta_k(1),...,\theta_k(M_k)\right) \tag{4.59}$$

of M_k measurements. Let

$$\boldsymbol{\theta} = \left(\theta_1,...,\theta_K\right) \tag{4.60}$$

be the vector of the K scans of measurements.

4.6.1.3 Associations

For each scan θ_k of measurements, there is an association function γ_k such that

$\gamma_k(m) =$ index of the target to which measurement m is associated

Define

$$\boldsymbol{\gamma} = \left(\gamma_1,...,\gamma_K\right) \tag{4.61}$$

to be the vector of association functions for the K scans. For the prior distribution on γ_k, we assume that $\gamma_k(m)$, for $m = 1,...,M_k$, are independent identically distributed random variables such that

$$\Pr\{\gamma_k(m) = n\} = \pi_{kn} > 0 \text{ for } n = 1,...,N \tag{4.62}$$

The probabilities π_{kn} are often chosen to be equal to $1/N$ if no prior information is available. In practice, we have not found their choice to be important unless they are chosen very near to one or zero.

Each association function is a vector of random variables giving the (unknown) associations for each scan. We have taken a Bayesian point of view and quantified the uncertainty about the association of measurements to targets by a (prior) probability distribution. Let Γ be the random vector representing the (unknown) association function γ. Then the prior on Γ is given by

$$\Pr\{\Gamma = \gamma\} = \prod_{k=1}^{K}\prod_{m=1}^{M_k}\pi_{k\gamma_k(m)} \tag{4.63}$$

Let $n_k = (n_{k1},...,n_{kM_k})$ denote a realization of the random vector $\gamma_k = (\gamma_k(1),...,\gamma_k(M_k))$. By this we mean $\gamma_k(m) = n_{km}$ for $m = 1,...,M_k$. When we write $\gamma = (n_1,...,n_K)$, we mean

$$\gamma_k(m) = n_{km} \text{ for } m = 1,\ldots.M_k \text{ and } k = 1,\ldots,K$$

4.6.1.4 Likelihood Functions

We now calculate the measurement, scan, multiscan, and conditional multiscan likelihood functions.

Measurement Likelihood

Let $\Theta_k(m)$ be the random variable representing the mth measurement in scan k. The likelihood function for the measurement $\theta_k(m)$ given that it is generated by target n is

$$l\big(\theta_k(m)\,|\,x_n(k)\big) = \Pr\big\{\Theta_k(m) = \theta_k(m)\,|\, X_n(t_k) = x_n(k)\big\}$$

The likelihood function for the measurement $\Theta_k(m) = \theta_k(m)$ given that $X(t_k) = x(k)$ is

$$\begin{aligned}
L_m\big(\theta_k(m)\,|\,x(k)\big) &= \Pr\big\{\Theta_k(m) = \theta_k(m)\,|\,X(t_k) = x(k)\big\} \\
&= \sum_{n=1}^{N} \Pr\big\{\gamma_k(m) = n\big\} l\big(\theta_k(m)\,|\,x_n(k)\big) \\
&= \sum_{n=1}^{N} \pi_{kn} l\big(\theta_k(m)\,|\,x_n(k)\big)
\end{aligned}$$

Scan Likelihood

Let $\Theta_k = (\Theta_k(1),\ldots,\Theta_k(M_k))$ and define the scan likelihood function

$$\begin{aligned}
L\big(\theta_k\,|\,x(k)\big) &= \Pr\big\{\Theta_k = \theta_k\,|\,X(t_k) = x(k)\big\} \\
&= \prod_{m=1}^{M_k} L_m\big(\theta_k(m)\,|\,x(k)\big) = \prod_{m=1}^{M_k}\sum_{n=1}^{N} \pi_{kn} l\big(\theta_k(m)\,|\,x_n(k)\big)
\end{aligned}$$

Multiscan Likelihood

Let $\Theta = (\Theta_1,\ldots,\Theta_K)$. Then

$$\begin{aligned}
\Pr\big\{\Theta = \theta\,|\,X = x\big\} &= \prod_{k=1}^{K} L\big(\theta_k\,|\,x(k)\big) \\
&= \prod_{k=1}^{K}\prod_{m=1}^{M_k}\sum_{n=1}^{N} \pi_{kn} l\big(\theta_k(m)\,|\,x_n(k)\big)
\end{aligned} \tag{4.64}$$

Equation (4.64) gives the likelihood for the K scans of data given a specified path for each of the N targets.

Conditional Multiscan Likelihood

The multiscan likelihood conditioned on an association γ is computed by

$$\Pr\{\Theta = \theta \mid \mathbf{X} = \mathbf{x}, \Gamma = \gamma\} = \prod_{k=1}^{K}\prod_{m=1}^{M_k} l\left(\theta_k(m) \mid x_{\gamma_k(m)}(k)\right) \qquad (4.65)$$

4.6.2 Posterior Distribution on Associations

In this section we compute the posterior distribution on associations given the measurements Θ and the target paths $\mathbf{x}$. Define

$$P(\theta, \mathbf{x}, \gamma) \equiv \Pr\{\Theta = \theta, \mathbf{X} = \mathbf{x}, \Gamma = \gamma\}$$
$$P(\theta \mid \mathbf{x}, \gamma) \equiv \Pr\{\Theta = \theta \mid \mathbf{X} = \mathbf{x}, \Gamma = \gamma\}$$

with similar definitions holding for other combinations and permutations of the variables $\theta, \mathbf{x}$, and γ.

Since $\mathbf{X}$ and Γ are independent,

$$P(\theta, \mathbf{x}, \gamma) = P(\mathbf{x})P(\gamma)P(\theta \mid \mathbf{x}, \gamma) \qquad (4.66)$$

By (4.57), (4.63), and (4.65),

$$P(\theta, \mathbf{x}, \gamma)$$
$$= \prod_{n=1}^{N}\left[q(x_n(1))\prod_{k=2}^{K}q_k\left(x_n(k) \mid x_n(k-1)\right)\right]\left[\prod_{k=1}^{K}\prod_{m=1}^{M_k}\pi_{kn_{km}} l\left(\theta_k(m) \mid x_{n_{km}}(k)\right)\right] \qquad (4.67)$$

where $\gamma = (n_1, \ldots, n_K)$. Since $P(\theta, \mathbf{x}) = P(\mathbf{x})\Pr\{\Theta = \theta \mid \mathbf{X} = \mathbf{x}\}$, we have from (4.64)

$$P(\theta, \mathbf{x})$$
$$= \prod_{n=1}^{N}\left[q(x_n(1))\prod_{k=2}^{K}q_k\left(x_n(k) \mid x_n(k-1)\right)\right]\left[\prod_{k=1}^{K}\prod_{m=1}^{M_k}\sum_{n=1}^{N}\pi_{kn} l\left(\theta_k(m) \mid x_n(k)\right)\right] \qquad (4.68)$$

From (4.67) and (4.68) we compute the posterior probability that $\Gamma = \gamma = (n_1, \ldots, n_K)$ by

$$P(\gamma \mid \theta, \mathbf{x}) = \frac{P(\theta, \mathbf{x}, \gamma)}{P(\theta, \mathbf{x})} = \prod_{k=1}^{K} \prod_{m=1}^{M_k} \frac{\pi_{k n_{km}} l\left(\theta_k(m) \mid x_{n_{km}}(k)\right)}{\displaystyle\sum_{n=1}^{N} \pi_{kn} l\left(\theta_k(m) \mid x_n(k)\right)}$$

Let

$$w(n, k, m \mid \mathbf{x}) = \frac{\pi_{kn} l\left(\theta_k(m) \mid x_n(k)\right)}{\displaystyle\sum_{n'=1}^{N} \pi_{kn'} l\left(\theta_k(m) \mid x_{n'}(k)\right)} \tag{4.69}$$

for $k = 1, \ldots, K$, $m = 1, \ldots. M_k$. Then $w(n, k, m \mid \mathbf{x}) = \Pr\{\gamma_k(m) = n \mid \theta, \mathbf{x}\}$ and

$$P(\gamma \mid \theta, \mathbf{x}) = \prod_{k=1}^{K} \prod_{m=1}^{M_k} w\left(\gamma_k(m), k, m \mid \mathbf{x}\right) \text{ for } \gamma = (n_1, \ldots, n_K) \tag{4.70}$$

Note that the posterior probabilities $w(n, k, m \mid \mathbf{x})$ for time t_k do not depend on measurements or target states at times other than t_k.

The sum of the right-hand side of (4.70) over all associations $\gamma = (n_1, \ldots, n_K) \in \Gamma$ equals one. That is

$$\sum_{(n_1, \ldots, n_K)} \prod_{k=1}^{K} \prod_{m=1}^{M_k} w\left(n_{km}, k, m \mid \mathbf{x}\right) = 1 \tag{4.71}$$

Let $(n_1, \ldots, n_K) \setminus n_{k'm'}$ indicate all $\gamma = (n_1, \ldots, n_K)$ for which $\gamma_{k'}(m') = n_{k'm'}$. Then

$$\sum_{(n_1, \ldots, n_K) \setminus n_{k'm'}} \prod_{k=1}^{K} \prod_{m=1}^{M_k} w\left(n_{km}, k, m \mid \mathbf{x}\right) = w\left(n_{k'm'}, k', m' \mid \mathbf{x}\right) \tag{4.72}$$

4.6.3 Expectation Maximization

We wish to find the set of target paths $\mathbf{x}^* = (x_1^*, \ldots x_N^*)$ that maximizes $P(\mathbf{x} \mid \theta)$, the posterior probability of the target paths $\mathbf{x}$ given the measurements θ. This is the MAP estimator for the target paths. One major difficulty in finding the MAP paths is that we do not know the correct association of measurements to targets. The expectation maximization (EM) method employed below provides a way out of this difficulty through the use of an auxiliary function defined in terms of an expectation over the posterior distribution on associations. In standard EM terminology, the unknown correct associations of measurements to targets are called "missing" variables. We employ a more Bayesian viewpoint wherein the unknown variables are represented by random variables with a specified prior distribution.

The EM method is a standard statistical technique developed in [18]. It is often applied in situations where finding the maximum likelihood (or MAP) estimator would be straightforward if some additional data were known. Applying a sequence of expectations over this unknown data and subsequent maximizations provides a method for finding the desired maximum likelihood estimator. In the discussion below we describe the application of the EM algorithm to the PMHT problem. Reference [19] provides more information about the theory and applications of EM algorithms.

4.6.3.1 Auxiliary Function

For a fixed path $\mathbf{x}'$, define the auxiliary function Q defined on path $\mathbf{x}$ as follows.

$$Q(\mathbf{x}\mid\mathbf{x}') = \sum_{\gamma} P(\gamma\mid\mathbf{x}',\theta)\ln P(\theta,\mathbf{x},\gamma) \tag{4.73}$$

Since $P(\theta,\mathbf{x},\gamma) = cP(\mathbf{x},\gamma\mid\theta)$, where $c = P(\theta)$ does not depend on $\mathbf{x}$ or γ, we have

$$Q(\mathbf{x}\mid\mathbf{x}') = \ln c + \sum_{\gamma} P(\gamma\mid\mathbf{x}',\theta)\ln P(\mathbf{x},\gamma\mid\theta)$$

and since $P(\mathbf{x},\gamma\mid\theta) = P(\mathbf{x}\mid\theta)P(\gamma\mid\mathbf{x},\theta)$, it follows that

$$Q(\mathbf{x}\mid\mathbf{x}') = \ln c + \ln P(\mathbf{x}\mid\theta) + \sum_{\gamma} P(\gamma\mid\mathbf{x}',\theta)\ln P(\gamma\mid\mathbf{x},\theta) \tag{4.74}$$

Let $\mathbf{P}$ be the set of probability distributions on a finite set $\{1,\dots,J\}$ where J is an arbitrary positive integer. Let $p = (p_1,\dots,p_J)$ denote a probability distribution in $\mathbf{P}$ and let $p' \in \mathbf{P}$ be fixed distribution in $\mathbf{P}$. By a straightforward Lagrange multiplier argument, one can show that

$$\max_{p\in\mathbf{p}} \sum_{j=1}^{J} \ln(p_j)p_j' = \sum_{j=1}^{J} \ln(p_j')p_j' . \tag{4.75}$$

From (4.74) and (4.75) it follows that if we find $\mathbf{x}''$ such that

$$Q(\mathbf{x}''\mid\mathbf{x}') > Q(\mathbf{x}'\mid\mathbf{x}') \tag{4.76}$$

then we must have

$$P(\mathbf{x}'' \mid \boldsymbol{\theta}) > P(\mathbf{x}' \mid \boldsymbol{\theta}) \tag{4.77}$$

One way to obtain to obtain an $\mathbf{x}''$ satisfying (4.76) is to find the $\mathbf{x}^*$ that maximizes $Q(\mathbf{x} \mid \mathbf{x}')$. Then setting $x'' = x^*$, we have $P(\mathbf{x}'' \mid \boldsymbol{\theta}) > P(\mathbf{x}' \mid \boldsymbol{\theta})$ unless equality holds. If equality holds, we have reached a stationary point that, except for unusual circumstances, will be a local maximum. The use of an auxillary function to find an x'' that satifies (4.77) is a special case of a more general numerical technique called iterative majorization [20].

4.6.3.2 EM Recursion

The EM recursion proceeds by alternating expectation and maximization steps. Suppose that we have obtained a set of target paths $\mathbf{x}^j$ at the jth step of the recursion. Then we do the following

Expectation step. Compute the expectation

$$Q(\mathbf{x} \mid \mathbf{x}^j) = \sum_{\gamma} \ln P(\boldsymbol{\theta}, \mathbf{x}, \gamma) P(\gamma \mid \mathbf{x}^j, \boldsymbol{\theta}) \tag{4.78}$$

Maximization step. Find $\mathbf{x}^{j+1}$ such that

$$Q(\mathbf{x}^{j+1} \mid \mathbf{x}^j) = \max_{\mathbf{x}} Q(\mathbf{x} \mid \mathbf{x}^j) \tag{4.79}$$

At the end of the maximization step, one returns to the expectation step with $\mathbf{x}^{j+1}$ in place of $\mathbf{x}^j$. One can start the recursion with any feasible set of target paths. If the likelihood function for target paths given by (4.64) is bounded, [18] shows that this algorithm converges to a stationary point of this likelihood function. In most cases, this point will be a local maximum.

It is clear that the usefulness of EM depends on being able to calculate and optimize the auxiliary function Q. We shall see that for the PMHT formulation of multiple target tracking, it is feasible to do this for nonlinear tracking problems through the use of particle filters. In the linear-Gaussian case, the form of Q leads to an elegant recursion involving a Kalman smoother.

4.6.4 Nonlinear PMHT

We now turn to the question of calculating the auxiliary function and then optimizing it to find the MAP set of target paths.

4.6.4.1 Calculating the Auxiliary Function

From the definition of Q in (4.73) combined with (4.67), and (4.70), we have

$$Q(\mathbf{x}\,|\,\mathbf{x}') = \sum_{\gamma=(n_1,\ldots,n_K)} \prod_{k=1}^{K} \prod_{m=1}^{M_k} w\big(n_{km},k,m\,|\,\mathbf{x}'\big) \ln P\big(\theta,\mathbf{x},\gamma\big)$$

where

$$\ln P\big(\theta,\mathbf{x},\gamma\big) = \sum_{n=1}^{N}\left[\ln q\big(x_n(1)\big) + \sum_{K=2}^{K} \ln q_k\big(x_n(k)\,|\,x_n(k-1)\big)\right]$$
$$+\sum_{k=1}^{K}\sum_{m=1}^{M_k} \ln l\big(\theta_k(m)\,|\,x_{n_{km}}(k)\big) + \sum_{k=1}^{K}\sum_{m=1}^{M_k} \ln \pi_{kn_{km}}$$

Observe that $Q(\mathbf{x}\,|\,\mathbf{x}')$ may now be written as the sum of N separate sums where the nth sum depends only on target n. In particular

$$Q(\mathbf{x}\,|\,\mathbf{x}') = \sum_{n=1}^{N} Q_n(\mathbf{x}\,|\,\mathbf{x}') \qquad (4.80)$$

where

$$Q_n\big(x_n\,|\,x_n'\big) = \ln q\big(x_n(1)\big) + \sum_{K=2}^{K} \ln q_k\big(x_n(k)\,|\,x_n(k-1)\big)$$
$$+\sum_{k=1}^{K}\sum_{m=1}^{M_k} \ln l\big(\theta_k(m)\,|\,x_n(k)\big) w\big(n,k,m\,|\,\mathbf{x}'\big) \qquad (4.81)$$
$$+\sum_{k=1}^{K}\sum_{m=1}^{M_k} \ln \pi_{kn} w\big(n,k,m\,|\,\mathbf{x}'\big)$$

To obtain the first line on the right-hand side of (4.81) we have used (4.71). The next two lines are obtained from (4.72) as shown in Section 4.6.6.

From (4.80) and (4.81), we see that the problem of finding N target paths to jointly maximize $Q(\mathbf{x}\,|\,\mathbf{x}')$ separates into N independent problems of finding the target path x_n to maximize $Q_n(x_n\,|\,x_n')$ for $n=1,\ldots,N$. This separation means that the computational load of PMHT grows linearly with number of contacts which provides the computational advantage to using the PMHT approximation for solving multiple target tracking problems.

4.6.4.2 Maximizing Q_n

Let

$$Q_n^0\left(x_n \mid x_n'\right) = Q_n\left(x_n \mid x_n'\right) - \sum_{k=1}^{K}\sum_{m=1}^{M_k} \ln \pi_{kn} w\left(n, k, m \mid \mathbf{x}'\right)$$

Since the subtracted sum on the right-hand side of the above equation does not depend on x_n, we may maximize $Q_n\left(x_n \mid x_n'\right)$ by finding x_n to maximize $Q_n^0\left(x_n \mid x_n'\right)$. As a result, we can maximize $Q_n(x_n \mid x_n')$ by maximizing

$$G_n(x_n) \equiv e^{Q_n^0(x_n \mid x_n')}$$

$$= q\left(x_n(1)\right)\prod_{m=1}^{M_1}\left[l\left(\theta_1(m) \mid x_n(k)\right)\right]^{w(n,1,m \mid \mathbf{x}')} \tag{4.82}$$

$$\times \prod_{k=2}^{K} q_k\left(x_n(k) \mid x_n(k-1)\right)\prod_{m=1}^{M_k}\left[l\left(\theta_k(m) \mid x_n(k)\right)\right]^{w(n,k,m \mid \mathbf{x}')}$$

We may maximize G_n through the use of a particle filter and a set of synthetic measurements as follows. For each scan θ_k, compute a synthetic likelihood function

$$\xi_n\left(\theta_k \mid x_n(k)\right) = \prod_{m=1}^{M_k}\left[l\left(\theta_k(m) \mid x_n(k)\right)\right]^{w(n,k,m \mid \mathbf{x}')} \tag{4.83}$$

and write

$$G_n\left(x_n\right) = q\left(x_n(1)\right)\xi_n\left(\theta_1 \mid x_n(1)\right)\prod_{k=2}^{K} q_k\left(x_n(k) \mid x_n(k-1)\right)\xi_n\left(\theta_k \mid x_n(k)\right) \tag{4.84}$$

so that $G_n(x_n)$ is proportional to the posterior probability on path x_n given the measurements θ and the synthetic likelihood functions defined in (4.83). Note that the synthetic likelihood functions depend on the posterior association probabilities computed from $\mathbf{x}'$.

One can simulate a large number of paths for target n from the Markov motion model for target n. Then for each path x_n compute $G_n(x_n)$ from (4.84) and choose the path x_n^* that produces the maximum over the paths. Set $\mathbf{x}' = \mathbf{x}^* = (x_1^*, \ldots, x_N^*)$ and compute new posterior association probabilities $w(n, k, m \mid \mathbf{x}')$ from (4.69).

Alternatively, one can apply the SIR particle filter in the recursive fashion described in Section 3.3 using the synthetic likelihood function in (4.83) to compute the posterior distribution on the target states at time t_k for $k = 1, \ldots, K$ for target n independently of the other targets. After the information update at time t_K and before the resampling step in the recursion in Section 3.3.2, select the particle with the maximum posterior probability for target n. The path x_n^* corresponding to this particle is the estimated MAP path for target n. Doing this for $n = 1, \ldots, N$ produces $x^* = (x_1^*, \ldots, x_N^*)$, which is the set of target paths that approximately maximize the auxiliary function Q.

One can continue the iteration by computing Q_n as given above and using a particle filter method to find a new $x' = x*$. This continues until a stopping rule is satisfied. At that point a local (near) maximum has been achieved, and the resulting target paths are the approximate MAP solution to the PMHT formulation of the multiple target tracking problem. A particle filter implementation of PMHT is presented in [21].

4.6.5 Linear-Gaussian PMHT

We present the PMHT algorithm for the linear-Gaussian case. Recall that $\eta(\cdot, \mu, \Sigma)$ is the density function for a Gaussian distribution with mean μ and covariance Σ. For the linear-Gaussian case we have

$$
\begin{aligned}
q\big(x_n(1)\big) &= \eta\big(x_n(1), \mu_n, \Sigma_n\big) \\
q_k\big(x_n(k) \mid x_n(k-1)\big) &= \eta\big(x_n(k), F_{kn} x_n(k-1), Q_{kn}\big) \\
l\big(\theta_k(m) \mid x_n(k)\big) &= \eta\big(\theta_k(m), H_{kn} x_n(k), R_{kn}\big)
\end{aligned}
\tag{4.85}
$$

where F_{kn}, H_{kn} are matrices of the appropriate dimensions and Q_{kn} and R_{kn} are covariance matrices. One can show (see [12] or [15]) that ξ in (4.83) is given by

$$
\xi_n\big(\theta_k \mid x_n(k)\big) = \eta\big(\tilde{\theta}_k^n, H_{kn} x_n(k), \tilde{R}_{kn}\big)
$$

where

$$
\tilde{\theta}_k^n = \frac{\sum_{m=1}^{M_k} w(n, k, m \mid x') \theta_k(m)}{\sum_{m=1}^{M_k} w(n, k, m \mid x')} \quad \text{and} \quad \tilde{R}_{kn} = \frac{R_{kn}}{\sum_{m=1}^{M_k} w(n, k, m \mid x')}
\tag{4.86}
$$

Thus (4.84) becomes

$$G_n(x_n) = \eta\big(x_n(1), \mu_n, \Sigma_n\big)\eta\big(\tilde{\theta}_1, H_{1n}x_n(1), \tilde{R}_{1n}\big)$$

$$\times \prod_{k=2}^{K} \eta\big(x_n(k), F_{kn}x_n(k-1), Q_{kn}\big)\eta\big(\tilde{\theta}_k, H_{kn}x_n(k), \tilde{R}_{kn}\big) \tag{4.87}$$

Finding a path x_n^* to maximize (4.87) is equivalent to finding the maximum likelihood solution to a Kalman smoothing problem. Since the mean path of the smoothed Kalman solution is the maximum likelihood path, one can find x_n^* by using one of the Kalman smoothing algorithms described in Section 3.2.3. The mean of that solution will be the desired x_n^*.

Setting $\mathbf{x}' = (x_1^*, \ldots, x_N^*)$, one then computes the new posterior association probabilities $w(n, k, m \mid \mathbf{x}')$ from (4.69) using the definitions in (4.85). From these one recomputes $\tilde{\theta}_k^n$ and $\tilde{R}_{kn}$ from (4.86) and uses these to redefine G_n for $n = 1, \ldots, N$. The solution from the Kalman smoother provides the updated set of target paths for each target. This recursion continues until a stopping rule is triggered. The resulting set of targets paths is an approximate MAP solution. This yields the linear-Gaussian PMHT algorithm given below.

4.6.6 Proof of (4.81)

We obtain the third line in (4.81) by evaluating

$$\sum_{\gamma=(n_1,\ldots,n_K)} \prod_{k=1}^{K}\prod_{m=1}^{M_k} w\big(n_{km}, k, m \mid \mathbf{x}'\big)\sum_{k'=1}^{K}\sum_{m'=1}^{M_{k'}} \ln \pi_{k'n_{k'm'}}$$

$$= \sum_{k'=1}^{K}\sum_{m'=1}^{M_{k'}}\sum_{\gamma=(n_1,\ldots,n_K)} \prod_{k=1}^{K}\prod_{m=1}^{M_k} w\big(n_{km}, k, m \mid \mathbf{x}'\big)\ln \pi_{k'n_{k'm'}}$$

$$= \sum_{k'=1}^{K}\sum_{m'}^{M_{k'}}\sum_{n_{k'm'}=1}^{N} \ln \pi_{k'n_{k'm'}} \sum_{\gamma=(n_1,\ldots,n_K)\setminus n_{k'm'}} \prod_{k=1}^{K}\prod_{m=1}^{M_k} w\big(n_{km}, k, m \mid \mathbf{x}'\big)$$

$$= \sum_{k'=1}^{K}\sum_{m'}^{M_{k'}}\sum_{n_{k'm'}=1}^{N} \ln \pi_{k'n_{k'm'}} w\big(n_{k'm'}, k', m' \mid \mathbf{x}'\big)$$

$$= \sum_{n_{k'm'}=1}^{N}\sum_{k'=1}^{K}\sum_{m'}^{M_{k'}} \ln \pi_{k'n_{k'm'}} w\big(n_{k'm'}, k', m' \mid \mathbf{x}'\big)$$

By an analogous computation with $\ln l\big(\theta_k(m) \mid x_{n_{km}}(k)\big)$ in place of $\ln \pi_{kn_{km}}$ we obtain the second line in (4.81).

Linear-Gaussian PMHT Algorithm

There are N targets and a scan of measurements $\theta_k = \left(\theta_k(1),\ldots,\theta_k(M_k)\right)$ at time t_k for $k = 1,\ldots,K$. Let

$$\boldsymbol{\theta} = \left(\theta_1,\ldots,\theta_K\right)$$

Specify the prior measurement to target association probabilities $\pi_{kn} > 0$ for $k = 1,\ldots,K$ and $n = 1,\ldots,N$.

For target n the initial state distribution at time t_1, the Markov transition functions, and the likelihood functions for the measurements are given by (4.85). Choose an initial path $x_n^{(0)} = \left(x_n^{(0)}(1),\ldots,x_n^{(0)}(K)\right)$ for target n. This path can be the mean of the prior motion model. Let

$$\mathbf{x}^{(0)} = \left(x_1^{(0)},\ldots,x_N^{(0)}\right)$$

Iteration.

Suppose that we have $\mathbf{x}^{(i)} = \left(x_1^{(i)},\ldots,x_N^{(i)}\right)$ for $i \geq 0$. For $n = 1,\ldots,N$, compute

$$w\left(n,k,m \mid \mathbf{x}^{(i)}\right) = \frac{\pi_{kn} l\left(\theta_k(m) \mid x_n^{(i)}(k)\right)}{\sum\limits_{n'=1}^{N} \pi_{kn'} l\left(\theta_k(m) \mid x_{n'}^{(i)}(k)\right)}$$

$$\tag{4.88}$$

$$\tilde{\theta}_k^n = \frac{\sum_{m=1}^{M_k} w\left(n,k,m \mid \mathbf{x}^{(i)}\right)\theta_k(m)}{\sum_{m=1}^{M_k} w\left(n,k,m \mid \mathbf{x}^{(i)}\right)} \quad ; \quad \tilde{R}_{kn} = \frac{R_{kn}}{\sum_{m=1}^{M_k} w\left(n,k,m \mid \mathbf{x}^{(i)}\right)}$$

Apply a Kalman smoother to find the mean path $\overline{x}_n = \left(\overline{x}_n(1),\ldots,x_n(K)\right)$ of target n given the measurements $\tilde{\theta}_k^n$ and Gaussian measurement likelihood functions

$$L_{kn}\left(\tilde{\theta}_k^n \mid x\right) = \eta\left(\tilde{\theta}_k^n, H_{kn}x, \tilde{R}_{kn}\right) \text{ for } n = 1,\ldots,N$$

and set

$$\mathbf{x}^{i+1} = \left(\overline{x}_1,\ldots,\overline{x}_N\right)$$

$$\tag{4.89}$$

Stopping Rule.

The iteration continues until a stopping rule is met. For example stop when

$$P\left(\boldsymbol{\theta},\mathbf{x}^{i+1}\right) - P\left(\boldsymbol{\theta},\mathbf{x}^i\right) < \varepsilon$$

for some $\varepsilon > 0$ where $P(\boldsymbol{\theta},\mathbf{x})$ is calculated from (4.68). The set of paths $\mathbf{x}^{j+1}$ is the approximate PMHT solution for the N target paths.

4.7 SUMMARY

Section 4.1 defines a general form of the multiple target tracking problem by extending the motion models and likelihood functions defined in Chapter 3 for single targets to multiple target motion models and multiple target measurement likelihood functions. Using these notions, we derived the Bayesian recursion for multiple target tracking. While elegant and general, this recursion is difficult to use when solving multiple target tracking problems because it requires the maintenance of target distributions on large joint state spaces.

Section 4.2 considers the special case in which measurements arrive in scans. For this case we developed the general case of Bayesian MHT. A major difficulty in multiple target tracking is to decide which contacts are associated to a target and which are false measurements. MHT deals with this problem by forming association hypotheses. Each hypothesis divides the set of contacts into subsets, one subset for the contacts associated to each target and one for false measurements. For each hypothesis, one calculates the distribution on target states using the associations specified by the hypothesis. The full joint distribution on target states is obtained by the mixture of the distributions resulting from each association hypothesis weighted by the probability of the hypothesis being correct.

The approach in Section 4.2 is of limited utility unless the target state distributions given a hypothesis are independent. In this case the joint target distribution is the product of independent single target distributions. In Section 4.3 we find conditions that guarantee this independence, and under these conditions prove the independent MHT recursion for the nonlinear non-Gaussian case. In Section 4.4, we specialize this recursion to the linear-Gaussian case. The linear-Gaussian MHT is the one most often used. However, a major difficulty with MHT is that the number of association hypotheses grows exponentially with the number of contacts. In order to keep the computational load bounded and reasonable, MHT algorithms approximate the full Bayesian solution by keeping only the highest probability hypotheses. Typically an upper bound on the number of hypotheses is established. If the number of hypotheses exceeds that bound, low-probability hypotheses are discarded until the bound is met. In addition, most MHT systems display the target state distributions for only the highest probability association hypothesis.

The MHT approximation described above produces excellent multiple target tracking systems when the linear Gaussian assumptions hold (at least approximately), but the software to implement these systems is complex. Typically a lot of computational effort is devoted to hypotheses that are minor variants of the highest probability hypothesis. Section 4.5 considers another approximation called JPDA. JPDA operates recursively and makes the following crucial approximations. (1) The target state distributions are independent. (2) The target state distributions updated for the measurements prior to time t_k and motion updated to the time t_k of the kth scan are a sufficient statistic for computing the posterior distributions given the measurements in the kth scan. (3) JPDA relaxes the hard-association assumption and allows soft associations, ones where the probability of a contact being associated to a target is equal to p where $0 \leq p \leq 1$. When JPDA updates the target distribution for the effect of this contact, it approximates the posterior target state distribution as the weighted mixture of the distribution updated with the contact information and the distribution not updated with the contact information using the weights p and $1 - p$. As originally developed in [2, 12], JPDA was applied using linear-Gaussian assumptions and Kalman filters to perform the tracking. Section 4.5 presents a nonlinear non-Gaussian version of JPDA that can be implemented using particle filters.

Section 4.6 considers a third approximation called PMHT. PMHT applies to scans of data but it relaxes the scan assumptions of MHT in two ways. First it allows for soft associations, and second it allows more than one contact per scan to be associated to a target. PMHT assumes that the number of targets N is known and fixed throughout the interval $[0, T]$ during which scans are received. In addition, PMHT finds the set of target tracks having the MAP but does not provide target state distributions. The advantage of PMHT is that the computational load grows linearly with the number of contacts and targets. Section 4.6 derives a nonlinear version of PMHT that can be implemented by particle filters and presents the more widely known version that holds under the linear-Gaussian assumptions.

Table 4.1 summarizes the assumptions required and outputs produced for the multiple target tracking methods described in this chapter. For all of these methods, the assumptions stated in Section 4.1 are assumed to hold.

4.8 NOTES

MHT was originally developed by Reid [9] for the linear-Gaussian case. Mori et al. presented a more general model of MHT in [22]. The model of MHT developed in [22] differs from the one presented in this chapter in a number of respects. It is more general in that it allows for an unbounded number of targets. To do this it represents the system state space as a union over n, the number of targets present, of target state spaces. However, it is more restrictive in an important way. The results apply only to situations in which all targets are identical. In our terminology they all have the same characteristics and same prior distribution on the target state space. The results in this chapter do not require this assumption.

Table 4.1

Summary of Multiple Target Tracking Methods, Outputs, and Assumptions

Method/Outputs	*Assumptions*
MHT Joint posterior target state distribution conditioned on each data association hypothesis. Probability each data association hypothesis is correct.	Observations are in the form of contacts as defined in Section 4.2.1. Contacts arrive in scans as defined in Section 4.2.2.
Independent MHT Posterior single-target state distributions that are mutually independent conditioned on a data association hypothesis.	MHT assumptions plus: The prior target motion stochastic processes are mutually independent. The scan association likelihood functions are conditionally deterministic as defined in Section 4.3.1.
Linear-Gaussian MHT Posterior single-target state distributions that are Gaussian and mutually independent condition on a data association hypothesis.	Independent MHT assumptions plus: Prior target motion processes are Gaussian. Contacts are elliptical as defined in Section 4.2.1.
JPDA A single posterior joint target state distribution that is the product of single-target state distributions. Each single-target state distribution is a mixture of the posterior distributions resulting from associating each contact from the present scan to the target plus the distribution resulting from associating no contacts to it.	Independent MHT assumptions plus: The present posterior joint target state distribution is a sufficient statistic for computing joint target state distributions based on future contacts. Contacts arrive in scans. Association of a measurement to a target is allowed to be soft, that is, a measurement can be associated to a target with probability p where $0 \leq p \leq 1$.
PMHT The (joint) maximum a posteriori probability set of tracks for the N targets over the time interval $[0, T]$	The number N of targets is fixed and known. The prior target motion stochastic processes are mutually independent. The scan assumption is relaxed to allow more than one measurement to be associated to a single target. The association process is modeled probabilistically. There is a (prior) probability distribution on the N targets. Each measurement is assigned to a target by an independent draw from this distribution. Conditioned on a set of measurement to target assignments, the measurement likelihood functions are mutually independent.
Linear-Gaussian PMHT Same output as PMHT.	PMHT assumptions plus: Prior target motion processes are Gaussian. Contacts are elliptical as defined in Section 4.2.1.

In the model given in this chapter, we require a bound $\overline{N}$ on the number of possible targets. This can be as high as we wish, so we do not view the requirement for an upper bound to be a substantial restriction. In turn, this allows a considerable simplification in the model and the results. Also, the notion of association hypothesis that we use seems to us simpler and easier to understand than the concept in [22].

Poore [8] has developed a method for selecting data association hypotheses that is related to MHT and applies to a special class of independent MHT situations. The method is a batch assignment procedure with a sliding window that uses integer programming to find the data association hypothesis with the highest probability of being correct. A scan window of size N is chosen. Using the tracks present at the beginning of the window and the contacts in the window, [8] finds the data association hypothesis with the highest probability of being correct and creates the resulting set of tracks. Then $K \geq 1$ new scans are received and R old scans are retained to form a new window of $N = K + R$ scans, and the process is repeated. So long as $K < N$, this method allows contacts to be reassigned to new tracks on the basis of new data until the contacts fall outside the sliding window. This is an approximate method of finding the maximum likelihood data association hypothesis. This is different than MHT because (in theory) MHT retains all the data association hypotheses and computes the tracks and target state distributions associated with each of them. In practice, MHT algorithms must limit the number of hypotheses that they carry, and they typically display the tracks (state estimates) for only the highest probability hypothesis. Thus, in practice, the difference between the results of the two methods may be small. This is particularly true for problems where the contacts have small measurement errors, the data rate is high, and the probability of not detecting a target during a scan is low. In this case, the effect of contacts in the "older" scans on the present track (state) estimates is small.

There have been many approaches to employing particle filters to perform nonlinear multiple target tracking. Of particular interest is a particle filter approach developed in [13,14] that implements the general multiple target recursion in (4.6)–(4.9). The sensor involved receives responses from a set of cells. The distribution of the response in a cell depends on the number of targets in the cell and is independent from cell to cell. Particles represent the number of targets present and the state of each target present. At each time period and for each particle, the filter computes the multiple target likelihood in Section 4.1.2 for the sensor response and applies it to the particle to compute the posterior distribution on the number and state of targets present. A motion model "moves" the particles to their states at the time of the next measurement, and the process is repeated, thereby implementing the general Bayesian multiple target recursion without contacts or association. The filter uses a Gaussian motion model for targets, but that could be replaced by a more general one.

References

[1]   Antony, R. T., *Principles of Data Fusion Automation,* Norwood, MA: Artech House, 1995.

[2]   Bar-Shalom, Y., and T. E. Fortman, *Tracking and Data Association,* New York: Academic Press, 1988.

[3]   Bar-Shalom, Y., and X. R. Li, *Multitarget-Multisensor Tracking: Principles and Techniques,* Storrs, CT: YBS Publishing, 1995.

[4]   Blackman, S. S., *Multiple Target Tracking with Radar Applications,* Norwood, MA: Artech House, 1986.

[5]   Goodman, I. R., R. P. S. Mahler, and H. T. Nguyen, *Mathematics of Data Fusion,* Boston: Kluwer Academic Publishers, 1997.

[6]   Mahler, R. P. S., *Statistical Multisource-Multitarget Information Fusion,* Norwood, MA: Artech House, 2007.

[7]   Hall, D. L., *Mathematical Techniques in Multisensor Data Fusion,* Norwood, MA: Artech House, 1992.

[8]   Poore, A. B., "Multidimensional assignment formulation of data association problems arising from multitarget and multisensor tracking," *Computational Optimization and Applications,* Vol. 3, 1994, pp. 27–57.

[9]   Reid, D. B., "An algorithm for tracking multiple targets," *IEEE Transactions on Automatic Control,* Vol. AC-24, No. 6, 1979, pp. 843–854.

[10]   Waltz, E., and J. Llinas, *Multisensor Data Fusion,* Norwood, MA: Artech House, 1990.

[11]   Bar-Shalom, Y., X. R. Li., and T. Kirubarajan, *Estimation with Applications to Tracking and Navigation,* New York: John Wiley & Sons, 2001.

[12]   Bar-Shalom, Y., P. K. Willett, and X. Tian, *Tracking and Data Fusion: A Handbook of Algorithms,* Storrs, CT: YBS Publishing, 2011.

[13]   Kreucher, C., K. Kastella, and A. O. Hero "Multitarget tracking using the joint multitarget probability density," *IEEE Transactions on Aerospace and Electronic Systems,* Vol 41, No. 4, 2005, pp. 1396-1414.

[14]   Morelande, M. R., C. M. Kreucher, and K. Kastella, "A Bayesian approach to multiple target detection and tracking," *IEEE Transactions on Signal Processing,* Vol. 55, No. 5, 2007, pp. 1589-1604.

[15]   Streit, R. L., and T. E. Luginbuhl, *Probabilistic Multi-Hypothesis Tracking,* NUWC-NPT Technical Report 10428, Naval Undersea Warfare Center, Newport, RI: February 15, 1995.

[16]   Streit, R. L., *Poisson Point Processes,* New York: Springer, 2010.

[17]   Streit, R. L., and T. E. Luginbuhl "Maximum likelihood method for probabilistic multi-hypothesis tracking" SPIE International Symposium, Orlando FL, 5-7 April 1994, *Proceedings: Signal and Data Processing of Small Targets* 1994, Vol. 2235, pp. 394-405.

[18]   Dempster, A. P., N. M. Laird, and D. B. Rubin, "Maximum likelihood from incomplete data via the EM algorithm (with discussion)," *Journal of the Royal Statistical Society, series B,* Vol. 39, 1977, pp 1 – 38.

[19]   Redner, R. A., and H. F. Walker, "Mixture densities, maximum likelihood and the EM algorithm," *SIAM Review,* Vol. 26, No. 2, 1984, pp. 195-239.

[20] Heiser, W. J., "Convergent computation by iterative majorization," Chapter 8 in *Recent Advances in Descriptive Multivariate Analysis*, W. J. Krzanowski (ed.), New York: Oxford University Press, 1995.

[21] Davey, S. J., "Histogram PMHT with particles," *Proceedings of 14th International Conference on Information Fusion*, Chicago IL, July 5 – 8, 2011, pp 779-786.

[22] Mori, S., et al., "Tracking and classifying multiple targets without apriori identification," *IEEE Transactions on Automatic Control*, Vol. AC-31, No. 5, 1986, pp. 401–409.

Chapter 5

Multitarget Intensity Filters

Chapter 4 begins by presenting a very general Bayesian recursion for multiple target tracking. It then restricts its attention to situations in which the observations are in the form of contacts that are received in scans. In this case the main obstacle to successful tracking is associating contacts with targets. Multiple hypothesis tracking (MHT) is the theoretically correct Bayesian solution to the tracking problem in this situation. However, computational considerations require that approximations be made. A typical approximation is to limit the number of association hypotheses that MHT carries. Joint probabilistic data association (JPDA) and probabilistic multiple hypothesis tracking (PMHT) represent other approximations or variations on MHT. Each of these deals explicitly with the problem of associating contacts to targets and producing a target state estimate for each target.

By contrast, intensity filters seek to estimate the number of targets per unit state space. Estimating the number of targets that are present in a surveillance region is a difficult problem that involves more than merely counting the number of sensor measurements. Not only can sensors fail to detect targets that are present, but spurious clutter measurements (false alarms) can be superimposed with target-originated measurements. Moreover, targets can leave the surveillance region, and new ones arrive at any time without notice.

Estimating the number of targets per unit state space is very closely related to estimating the distribution of multitarget state; however, they are phenomenologically different models. Intensity filters do not explicitly compute associations (or association probabilities) of contacts to targets nor do they estimate target tracks. Instead they compute intensity functions that estimate the density (in target state space) of the number of targets. In Chapter 7, we will show how to combine multitarget intensity filtering with likelihood ratio tracking to estimate the number of targets present and produce track estimates for the identified targets.

We use finite point processes to develop intensity filters, and in keeping with the approach of this book, take a Bayesian point of view. Throughout this chapter

161

we use the finite point process terminology that is pervasive in both the mathematical and telecommunication literatures. Random finite set (RFS) terminology is often used to describe finite point processes in the comparatively smaller tracking literature. We have chosen to use the finite point process terminology to be consistent with the larger technical community. We will identify the equivalent RFS terminology when appropriate.

Organization of the Chapter

Section 5.1 defines nonhomogeneous Poisson point processes (PPPs). PPPs are fully characterized by their intensity functions, a property that makes them useful as approximations of more general finite point processes. They are used in this chapter as approximations of the Bayes posterior process, which, as will be seen, is a finite point process that is not Poisson. Transformations of PPPs are employed to model target motion, detections, measurements, and births and deaths of targets. Further discussion of PPPs and their applications can be found in [1].

Section 5.2 derives the intensity filter (iFilter) using Bayesian analysis and basic properties of PPPs. The derivation is very much in the spirit of Chapter 4; that is, the measurement-to-target assignments are treated explicitly (in the derivation not in the iFilter algorithm itself). The derivation is applied to the traditional single target state space S augmented with a null point ϕ that facilitates Markovian modeling of target birth and death. The Bayes posterior is seen to be a finite point process, but not a PPP. To incorporate new sensor scan information in a recursive manner, we approximate the Bayes posterior point process on $S^+ = S \cup \phi$ by a PPP.

The probability hypothesis density (PHD) filter was derived in [2, 3] on the target state space S. A PHD is the intensity function of a PPP that approximates the Bayes posterior point process on S. Section 5.3 shows how to recover the PHD filter on S from the iFilter derived on the augmented space S^+. Section 5.3 also gives a simple, but satisfying, intuitive interpretation of both filters.

Section 5.4 gives an alternative derivation of the iFilter and PHD filter using classical probability generating functions (PGFs) for discrete random variables. The methods of this and the remaining sections of this chapter are very different in spirit from those used elsewhere in the book, but they give more general results. The iFilter is derived on finite grids in the target and measurement spaces. Filters on finite grids are useful in applications involving data that are aggregated into histogram cell counts and so do not correspond to distinct point measurements. The small grid cell limit is taken to eliminate the grid; the limiting filter is seen to be the PHD filter for point measurements. The PGF approach reveals the combinatorial underpinnings of the filter and shows that the multitarget model is a counting model. In other words, the PHD filter is a counting filter whose goal is to estimate the number of targets per unit state space.

Section 5.5 discusses the problem of estimating the number of targets when the targets are extended in the sensor measurement space—that is, when targets can produce a random number of independent and identically distributed (i.i.d.) sensor measurements.

Section 5.6 provides a summary, but it is more than a summary. It discusses the phenomenology of the multitarget model, and why our efforts to understand it led to the organization of ideas in the present chapter. It also describes the advantages of the different derivations and methods.

Section 5.7 gives some notes on topics not included in this chapter, as well as some background material on methods and applications.

5.1 POINT PROCESS MODEL OF MULTITARGET STATE

The simplest and perhaps most natural way to define multitarget state is to say that it comprises the number of targets and a state (e.g., position and velocity) for each of the individual targets. One way to proceed from this definition is to assume a maximum number $\bar{N} < \infty$ of targets and then define the multitarget state to be the vector whose $\bar{N}$ components are the state vectors of the individual targets with the state ϕ indicating target not present. This is the approach taken in Chapter 4. Under the conditions of the independence theorem in Section 4.3, the conditional posterior distribution on multitarget state, given an association hypothesis, factors into a product of posteriors, one for each target. Generally, the unconditioned Bayes posterior does not factor, and it is necessary to evaluate it on the full state space. The methods of Chapter 4 are designed for this problem.

Multitarget state does not have to be represented by a multitarget state vector. For intensity filters, the notion of multitarget state as presented in the preceding paragraph is weakened in the following way. The multitarget state contains a state for each target, but knowledge of which state belongs to which target is not retained. The change means that the multitarget state is no longer a vector of target states but an ordered pair of the form

$$\left(n, \{x_1, ..., x_n\}\right) \tag{5.1}$$

where n is the number of targets and $\{x_1, ..., x_n\} \in \mathcal{E}(S)$, the set of all finite lists of points, or target states, in S. (Lists can have repeated elements. See Section 5.1.1.) Permuting the order of the x's does not change the meaning of the multitarget state in (5.1), but it does change the meaning of the multitarget state vector. Statisticians would say that targets are statistically identifiable in the multitarget vector but not in (5.1).

Using the state space $\mathcal{E}(S)$ changes the mathematical character of the random variables used to model the multitarget process. With a multitarget vector, the multitarget process has outcomes that are vectors in $(S \cup \phi)^{\bar{N}}$. In contrast, with state defined by (5.1) the multitarget process has outcomes in $\mathcal{E}(S)$.

A finite point process is a random variable whose outcomes are in the set of all finite lists of points in a given set. The advantage of viewing finite point processes as random variables is that we benefit immediately from the rich Bayesian theory that is available in the open literature. Various summary statistics for these processes have long been known and used in practice, and they can be adapted for use in multitarget tracking.

Multitarget Process

In this chapter we shall use PPPs as our model for the multitarget process. Sections 5.1.1–5.1.3 define PPPs and provide some basic properties. In the spirit of Chapters 3 and 4, we also specify a target motion model and a sensor measurement model to complete the definition of the multitarget process. The motion and measurement processes are defined in Sections 5.1.4 and 5.1.5. Thinning is considered in Section 5.1.6; thinning will be used to model target death. Section 5.1.7 discusses the extended, or augmented, state space used to derive the intensity filter.

5.1.1 Basic Properties of PPPs

Let Ξ be the random variable whose values are realizations of a PPP. Every realization $\Xi = \xi$ of a PPP on a bounded set $\mathcal{R}$ is comprised of a number $n \geq 0$ and n locations $\{x_1, \ldots, x_n\}$ of points in $\mathcal{R}$. The realization is denoted by the ordered pair $\xi = (n, \{x_1, \ldots, x_n\})$. The set notation signifies that the ordering of the points x_i is irrelevant, but not that the points are necessarily distinct. It is better to think of $\{x_1, \ldots, x_n\}$ as a list. From the definition below, it will be clear that the list is a set (i.e., there are no repeated elements) with probability one for continuous spaces S. When S also has discrete elements, however, the list can contain multiple copies of the discrete elements.

Including n in the realization ξ makes expectations easier to define and manipulate. If $n = 0$, then $\xi = (0, \varnothing)$, where $\varnothing$ is the empty set. The event space of the PPP is

$$\mathcal{E}(\mathcal{R}) = \{(0, \varnothing)\} \cup_{n=1}^{\infty} \left\{ (n, \{x_1, \ldots, x_n\}) : x_j \in \mathcal{R}, j = 1, \ldots, n \right\} \qquad (5.2)$$

This is also the event space of the Bayes posterior point process that is derived later in the chapter.

Every PPP on S is characterized by a function $\lambda(s)$, $s \in S$, called the intensity. If $\lambda(s) = c$ for all s, where $c \geq 0$ is a constant, the PPP is said to be homogeneous; otherwise, it is nonhomogeneous. It is assumed that $\lambda(s) \geq 0$ for all s and, for continuous spaces S,

$$0 \leq \int_{\mathcal{R}} \lambda(s)\, ds < \infty \tag{5.3}$$

for all bounded subsets $\mathcal{R}$ of S, that is, subsets that are contained in some finite radius sphere. The intensity $\lambda(s)$ need not be a continuous function (e.g., it can have step discontinuities). Bounded subsets are "windows" in which PPP realizations are observed. By stipulating a window we avoid issues with infinite sets; for example, realizations of homogeneous PPPs on $S = \mathbb{R}^m$ have an infinite number of points but only a finite number in any window.

The intensity function can be defined when a countable number of discrete elements are joined to S, and when the space comprises only a countable number of discrete elements (i.e., it has no continuous component). For further discussion of these cases, see Section 5.1.7.

Definition. A finite point process with intensity $\lambda(s)$, $s \in S$, is a PPP if and only if its realizations ξ on bounded subsets $\mathcal{R}$ are generated (simulated) by the following procedure.

If $\int_{\mathcal{R}} \lambda(s)\, ds = 0$, then $\xi = (0, \varnothing)$.

If $\int_{\mathcal{R}} \lambda(s)\, ds > 0$, the realization ξ is obtained as follows:

- Step 1. The number n of points is a sample from the discrete Poisson variate N with

$$p_N(n) \equiv \Pr\{N = n\} = \exp\left(-\int_{\mathcal{R}} \lambda(s)\, ds\right) \frac{\left(\int_{\mathcal{R}} \lambda(s)\, ds\right)^n}{n!} \tag{5.4}$$

The mean number of points is $E[N] = \int_{\mathcal{R}} \lambda(s)\, ds$.

- Step 2. The n points $x_j \in \mathcal{R}, j = 1, \ldots, n$, are obtained as i.i.d. samples of a random variable X on $\mathcal{R}$ with probability distribution function (pdf)

$$p_X(s) = \frac{\lambda(s)}{\int_{\mathcal{R}} \lambda(s)\, ds} \quad \text{for } s \in \mathcal{R} \tag{5.5}$$

The output of step 2 is the ordered n-tuple $\xi_o = (n, (x_1, \ldots, x_n))$.

- Step 3. The realization of the PPP is $\xi = (n, \{x_1, \ldots, x_n\})$; that is, the order in which the points were generated is "forgotten."

Note that the list $\{x_1, \ldots, x_n\}$ is a set of distinct elements with probability one if the intensity is an ordinary function and the space S is continuous. If the space S is discrete, or like the augmented space S^+ has discrete components, then the list $\{x_1, \ldots, x_n\}$ can have repeated elements.

5.1.2 Probability Distribution Function for a PPP

The generation procedure above defines the pdf for the random variable Ξ corresponding to the PPP. The pdf $p_\Xi(\xi)$ of the unordered realization ξ is

$$
\begin{aligned}
p_\Xi(\xi) &= p_N(n) \Pr\{\{x_1, \ldots, x_n\} \mid N = n\} \\
&= \exp\left(-\int_{\mathcal{R}} \lambda(s)\,ds\right) \frac{\left(\int_{\mathcal{R}} \lambda(s)\,ds\right)^n}{n!} \, n! \prod_{j=1}^{n} \frac{\lambda(x_j)}{\int_{\mathcal{R}} \lambda(s)\,ds} \\
&= \exp\left(-\int_{\mathcal{R}} \lambda(s)\,ds\right) \prod_{j=1}^{n} \lambda(x_j) \qquad\qquad \text{for } n \geq 0
\end{aligned}
\tag{5.6}
$$

The $n!$ in the numerator of (5.6) arises from the fact that there are $n!$ equally likely ordered i.i.d. trials that can generate the list $\{x_1, \ldots, x_n\}$. For the case $n = 0$ in (5.6), we follow the usual convention that a product from $j = 1$ to 0 is equal to one. The pdf of Ξ is parameterized by the intensity function λ. The expression (5.6) is used in estimation problems involving data sets for which data order is irrelevant. The pdf of the ordered realization ξ_o is $p_\Xi(\xi)/n!$.

Several useful and important properties of PPPs can be derived directly from the probability structure just defined. These will be presented below. Further discussions of the properties of PPPs are readily available (e.g., [1, 4]).

5.1.3 Superposition of Point Processes

PPPs are superimposed, or summed, if their realizations are combined into one event. Let Ξ and Υ denote two PPPs on S, and let their intensities be $\lambda(s)$ and $\nu(s)$. If $(n, \{x_1, \ldots, x_n\})$ and $(m, \{y_1, \ldots, y_m\})$ are realizations of Ξ and Υ, respectively, then the combined event is $(n + m, \{x_1, \ldots, x_n, y_1, \ldots, y_m\})$. The sum $\Xi + \Upsilon$ is a point process on S, but it is not necessarily a PPP unless Ξ and Υ are mutually independent. Knowledge of which points originate from which realization is assumed lost, and this fact is crucial to showing that the probability of the combined event is equal to that of a realization of this combined event for a PPP with intensity $\lambda(s) + \nu(s)$. More generally, the superposition of any number of mutually independent PPPs is a PPP, and its intensity is the sum of the intensities of the constituent PPPs. See Section 2.5 of [1].

5.1.4 Target Motion Process

Targets move according to a Markov process as in Chapters 3 and 4. A PPP whose points undergo Markov motion is still a PPP. Let q be the transition function for the Markov motion process, so that the probability that the point $x \in S$ transitions to $x' \in S$ is $q(x' \mid x)$. If $\xi = (n, \{x_1, \ldots, x_n\})$ is a realization of the PPP Ξ with intensity $\lambda(x)$, the realization after the Markov transition is $\eta \equiv (n, \{x_1', \ldots, x_n'\}) \in \mathcal{E}(S)$, where x_j' is a realization of the pdf $q(\cdot \mid x_j), j = 1, \ldots, n$. The transformed process, denoted $q(\Xi)$, is a PPP on S with intensity

$$\nu(x') = \int_S q(x' \mid x)\lambda(x)\,dx \tag{5.7}$$

A proof of this result is given in Section 5.2.8.

5.1.5 Sensor Measurement Process

The sensor measurement process maps a point $x \in S$ to a point (measurement) $y \in T$, where the space T can be different from S. Let Ξ be a target PPP on S. If $l(\cdot \mid x)$ denotes the measurement likelihood function for a measurement originating at $x \in S$, then the measurement process is a PPP and

$$\upsilon(y) = \int_S l(y \mid x)\lambda(x)\,dx, \; y \in T \tag{5.8}$$

is its intensity function. A proof of this result is given in Section 5.2.8.

5.1.6 Thinning a Process

Thinning a point process by probabilistically reducing the number of points in the realizations produces another point process. Let Ξ be a point process on S. For every $x \in S$, let $1 - \alpha(x)$, $0 \le \alpha(x) \le 1$, be the probability that the point x is removed from any realization that contains it. For the realization $\xi = (n, \{x_1, \ldots, x_n\})$, the point x_j is retained with probability $\alpha(x_j)$ and dropped with probability $1 - \alpha(x_j)$. Thinning of the points is mutually independent. The thinned realization is $\xi_\alpha = (m, \{x_1', \ldots, x_m'\})$, where $m \le n$ is the number of points that survive the thinning. Knowledge of the number n of points in ξ is assumed lost in the thinned process. This is called Bernoulli thinning because the thinned process is the result of a sequence of Bernoulli trials on the points in the realization ξ. If Ξ is a PPP, then the thinned process, denoted by Ξ_α, is a PPP and

$$\lambda_\alpha(x) = \alpha(x)\lambda(x) \qquad (5.9)$$

is its intensity. In addition, the points removed from the realizations of Ξ comprise a PPP, $\Xi_{1-\alpha}$, whose intensity is $\lambda_{1-\alpha}(x) = (1 - \alpha(x))\lambda(x)$. A proof of this result is given in Section 5.2.8.

It is surprising perhaps that the thinned processes Ξ_α and $\Xi_{1-\alpha}$ are independent. The coloring theorem [4, Chap. 6] generalizes this result to multiple thinned processes: If a point x in a realization of the PPP Ξ is colored (i.e., labeled) with one of r colors according to a discrete random variable with outcome probabilities $p_1(x),...,p_r(x)$ that sum to one and the coloring of x is independent of the coloring of x' for $x' \neq x$, then the point process Ξ_j corresponding to the jth color is a PPP with intensity $\lambda_j(x) = p_j(x)\lambda(x)$; moreover, these PPPs are mutually independent.

5.1.7 Augmented Spaces

Straightforward modifications are needed for PPPs on the augmented space $S^+ = S \cup \phi$. The intensity is defined for all $s \in S^+$. The intensity $\lambda(\phi)$ is dimensionless, unlike the values of $\lambda(s)$ for $s \in S$. The bounded sets of S^+ are $\mathcal{R}$ and $\mathcal{R}^+ \equiv \mathcal{R} \cup \phi$, where $\mathcal{R}$ is a bounded subset of S. Integrals of $\lambda(s)$ over bounded subsets of S^+ must be finite; thus, the requirement (5.3) holds and is supplemented with

$$0 \leq \int_{\mathcal{R}^+} \lambda(s)\, ds \equiv \lambda(\phi) + \int_{\mathcal{R}} \lambda(s)\, ds < \infty \qquad (5.10)$$

The event space is $\mathcal{E}(\mathcal{R}^+)$. The event space $\mathcal{E}(\mathcal{R})$ is a proper subset of $\mathcal{E}(\mathcal{R}^+)$.

The pdf of this PPP is unchanged, except that the integrals are over either $\mathcal{R}$ or $\mathcal{R}^+$, as the case may be. Superposition and thinning are unchanged. The intensity functions of the target motion and sensor measurement processes are also unchanged, except that the integrals are over S^+.

Other cases are treated similarly. When S is augmented with a countable number of discrete elements, say $\{\phi_i\}$, with corresponding intensities $\{\lambda(\phi_i)\}$, the term $\lambda(\phi)$ in (5.10) is replaced by the sum $\Sigma_i\, \lambda(\phi_i)$. When the space comprises only discrete elements, then the integral in (5.10) is omitted.

5.2 iFILTER

In this section we show that the multitarget intensity filter (iFilter) can be understood in elementary terms. We present a self-contained derivation from Bayesian principles, starting from the initial assumption that the multitarget process at the previous time step is a PPP with known intensity function. The Bayes information-updated multitarget process is another point process. The distribution of this posterior process is derived and shown not to be that of a PPP. However, all the single-target marginal distributions of the posterior are seen to be identical. This leads to a PPP approximation of the Bayes posterior process, and this approximation closes the Bayes recursion—we began the recursion with a PPP model of multitarget process, and we end with a model of the same mathematical form.

The derivation assumes familiarity with single target Bayesian filtering and with the properties of PPPs at the elementary level discussed in the previous section. The augmented state space S^+ plays a key role in the combinatorics of the measurement to target assignments. On S the intensity filter reduces to the PHD filter as discussed in Section 5.3.

The methods used in this section are consistent with the general point of view of Chapter 4. They reveal the Bayesian character of the filtering, as well as the combinatorial modeling of the assignment problem.

5.2.1 Augmented State Space Modeling

The auxiliary point ϕ needs interpretation in tracking applications. For single target tracking, it represents the "target absent" hypothesis. This interpretation works because of the assumed dichotomy—a single target is either present or not. The interpretation of ϕ in the multitarget setting is more complex because there are time-varying numbers of targets, each of which may or may not produce a measurement, and there are spurious clutter measurements. Clutter measurements can be produced by random fluctuations in the environment or in the sensor itself. More often clutter measurements arise from physical phenomena (such as objects on the ocean bottom in the case of active acoustic sensors) that produce responses that are difficult to distinguish from that of a real target. Over time clutter measurements tend to be kinematically inconsistent with target motion. By contrast (real) targets in S generate measurement sequences that are consistent with target motion over time.

For multitarget intensity filters we expand the role of ϕ to include the generation of clutter measurements. We do this by specifying an expected number $\lambda(\phi)$ of clutter "targets." Each clutter target has probability $P^D(\phi)$ of generating a clutter measurement. The clutter measurements are assumed to be a PPP on the measurement space with $P^D(\phi)\lambda(\phi)$ being the expected number of clutter measurements. This is an endogenous model of clutter. It is more common in the literature to superimpose spurious clutter data from an unmodeled source upon the target-originated measurements. This kind of "exogenous" clutter model is used in Section 5.4. While a priori there is no obvious reason to prefer one clutter model over the other (they are both somewhat artificial), the model we have chosen has the virtue of allowing us to compute the distribution of the Bayes posterior point process in a simple and straightforward fashion. Moreover, we show in Section 5.3 that the exogenous clutter model may be obtained as a special case of the endogenous one that we use.

The motion model on the augmented space S^+ is defined by the transition function $q(x \mid x')$, $x, x' \in S^+$. It is understood in the same way as in Chapter 4. For $x \in S$, the quantities $q(x|\phi)$ and $q(\phi|x)$ have natural interpretations as the probabilities of target birth and death, respectively. The transition function must be defined so that

$$\int_{S^+} q(x \mid x')\, dx = q(\phi \mid x') + \int_S q(x \mid x')\, dx = 1 \quad \text{for} \quad x' \in S^+ \qquad (5.11)$$

where the integral over S^+ is expanded using the analog of the definition (5.10).

The sensor measurement likelihood function must also be defined on S^+. For $x \in S$, $l(y|x)$ for $y \in T$ is the usual measurement likelihood function. For $x = \phi$, $l(y|\phi)$ is defined to be the probability that a clutter measurement has the value y. These definitions are the same as used in Chapter 4.

5.2.2 Predicted Detected and Undetected Target Processes

Denote the multitarget point process at the previous time step by Ξ, the motion updated multitarget process by Ξ^-, and the Bayesian information updated process by Ξ^+. These processes are defined on S^+. For the filters considered here, Ξ is assumed to be a PPP. Denote its intensity function by $f^\Xi(\cdot)$. Target motion from the previous time step to the current one is assumed to be Markovian.

For $x \in S^+$, let $d(x)$ denote the probability that a target at x does not survive the transition to the next time step. It is a thinning probability and does not need to integrate to one. Given the transition model $q(x|x')$ on the augmented space S^+, the motion-updated process Ξ^- is a PPP on S^+ and its intensity is

$$f^{\Xi^-}(x) = \int_{S^+} q\left(x \mid x'\right)\left(1 - d(x')\right) f^\Xi(x')\, dx' \quad \text{for} \quad x \in S^+ \qquad (5.12)$$

This expression is easily understood. Target survival is an independent thinning process with survival probability $1 - d(x)$. Using (5.9) we see that the surviving target process is a PPP with intensity $(1 - d(x)) f^\Xi(x)$. Target motion is a transformation of the form (5.7), so the motion-updated surviving multitarget process is a PPP with intensity (5.12).

The motion-updated target process is split by the detection process into two processes—detected and undetected. Let $P^D(x)$ denote the probability of detecting a target located at x. The detected target process, denoted by $\Xi^- D$, is the motion updated target process after thinning by $P^D(x)$, so it is a PPP. Using (5.9) we see that its intensity is

$$f^{\Xi^- D}(x) = P^D(x) f^{\Xi^-}(x) \quad \text{for } x \in S^+ \tag{5.13}$$

The undetected target process is denoted by $\Xi^- U$; its intensity is

$$f^{\Xi^- U}(x) = (1 - P^D(x)) f^{\Xi^-}(x) \quad \text{for } x \in S^+ \tag{5.14}$$

By the coloring theorem (with two colors), the detected and undetected target processes are mutually independent.

5.2.3 Measurement Process

The measurement process, denoted by Υ, is a PPP on T. The sensor likelihood function is $l(y \mid x)$. Detected targets are assumed to generate exactly one measurement. The intensity function for the measurement process is

$$
\begin{aligned}
\lambda(y) &= \int_{S^+} l(y \mid x) P^D(x) f^{\Xi^-}(x) \, dx \\
&= \lambda_\phi(y) + \int_S l(y \mid x) P^D(x) f^{\Xi^-}(x) \, dx
\end{aligned} \tag{5.15}
$$

where the clutter measurement intensity is

$$\lambda_\phi(y) = l(y \mid \phi) P^D(\phi) f^{\Xi^-}(\phi) \tag{5.16}$$

Since

$$\int_T l(y \mid x) \, dy = 1 \quad \text{for } x \in S^+$$

the expected number of measurements is equal to the expected number of detected targets; that is,

$$\int_T \lambda(y) \, dy = \int_{S^+} P^D(x) f^{\Xi^-}(x) \, dx = \int_{S^+} f^{\Xi^- D}(x) \, dx \tag{5.17}$$

This result follows from (5.13) and (5.15).

5.2.4 Bayes Posterior Point Process (Information Update)

Let $v_o = \left(m, (y_1, ..., y_m)\right)$ denote an ordered realization of the measurement process Υ at the current scan, where $m > 0$. (The special case $m = 0$ is discussed in Section 5.2.6.) Let $\xi_o = \left(n, (x_1, ..., x_n)\right)$ be an ordered realization of the multitarget detected process $\Xi^- D$. Then, using Bayes' theorem, the posterior distribution on ξ_o is given by

$$\Pr\{\xi_o \mid v_o\} = \Pr\{v_o \mid \xi_o\} \frac{\Pr\{\xi_o\}}{\Pr\{v_o\}} \tag{5.18}$$

We now compute the three functions on the right-hand side of (5.18).

The basic properties of the measurement to target assignment problem yield the conditional probability $\Pr\{v_o \mid \xi_o\}$ as follows:

$$\Pr\{v_o \mid \xi_o\} = \begin{cases} \dfrac{1}{m!} \displaystyle\sum_\sigma \prod_{j=1}^{m} l\left(y_{\sigma(j)} \mid x_j\right) & \text{for } m = n \\[4mm] 0 & \text{for } m \neq n \end{cases} \tag{5.19}$$

where the sum is over $\sigma \in \mathrm{Sym}(m)$, the set of all permutation on the first m positive integers. The sum over permutations reflects the fact that we assume a priori that all assignments of measurement to targets are equally likely. The division by $m!$ is needed because we are dealing with ordered realizations of ξ and v. (It makes the integral of the left-hand side of (5.19) over all possible values of the y's equal to one.) The reason that the likelihood function is zero for $m \neq n$ is the nature of the augmented multitarget state space—all measurements are accounted for by detected targets in S or ϕ.

Since the predicted multitarget and measurement processes are PPPs, we have from the pdf for ordered realizations that

$$\Pr\{\xi_o\} = \frac{1}{n!} P_{\Xi^- D}(\xi) = \frac{1}{n!} \exp\left(-\int_{S^+} f^{\Xi^- D}(s)\, ds\right) \prod_{j=1}^{n} f^{\Xi^- D}(x_j) \tag{5.20}$$

$$\Pr\{v_o\} = \frac{1}{m!} P_\Upsilon(v) = \frac{1}{m!} \exp\left(-\int_T \lambda(y)\, dy\right) \prod_{j=1}^{m} \lambda(y_j) \tag{5.21}$$

Substituting (5.19)–(5.21) into (5.18) and using the identity (5.17) gives

$$\Pr\{\xi_o \mid \upsilon_o\} = \begin{cases} \dfrac{1}{m!} \displaystyle\sum_\sigma \prod_{j=1}^m \dfrac{l\left(y_{\sigma(j)} \mid x_j\right) f^{\Xi^-D}(x_j)}{\lambda(y_{\sigma(j)})} & \text{for } m = n \\[4mm] 0 & \text{for } m \neq n \end{cases} \qquad (5.22)$$

The expression (5.22) clearly has a different mathematical form from the pdf (5.20) of an ordered realization of a PPP. Consequently, the Bayes posterior point process is a finite point process that is not a PPP. To close the Bayesian recursion, we approximate it with a PPP.

5.2.5 PPP Approximation

The Bayes posterior distribution (5.22) has the property that its single target marginal distributions are all equal. That is, integrating (5.22) over all the x's except x_ℓ (say) gives

$$\begin{aligned}
\Pr\{x_\ell \mid \upsilon_o\} &\equiv \int_{S^+} \cdots \int_{S^+} \Pr\{\xi_o \mid \upsilon_o\} \, dx_1 \cdots dx_{\ell-1} \, dx_{\ell+1} \cdots dx_m \\
&= \frac{1}{m!} \sum_{r=1}^m \sum_{\substack{\sigma \in \mathrm{Sym}(m) \\ \sigma(\ell)=r}} \frac{l\left(y_{\sigma(\ell)} \mid x_\ell\right) f^{\Xi^-D}(x_\ell)}{\lambda(y_{\sigma(\ell)})} \\
&= \frac{1}{m} \sum_{r=1}^m \frac{l\left(y_r \mid x_\ell\right) f^{\Xi^-D}(x_\ell)}{\lambda(y_r)}
\end{aligned} \qquad (5.23)$$

This identity holds for all $x_\ell \in S^+$.

The Bayes posterior density is approximated by the product of its single target marginal densities; that is,

$$\Pr\{x_1, \dots, x_m \mid \upsilon_o\} \cong \Pr\{x_1 \mid \upsilon_o\} \cdots \Pr\{x_m \mid \upsilon_o\} \qquad (5.24)$$

The product approximation is called the mean field approximation in the machine learning community, where it is widely used to factor multivariate pdfs into a product of marginal (or latent) variables to greatly reduce storage and facilitate subsequent analysis [6]. Given the product approximation, it is natural to approximate the Bayes posterior by a PPP whose intensity function is proportional to $\Pr\{x \mid \upsilon_o\}$, $x \in S^+$, where the constant of proportionality is the expected number of targets. Matching the expected number of targets to that of the Bayes posterior process gives the proportionality constant m, since the Bayes posterior always has exactly m targets due to the augmented space S^+. Hence the intensity of the PPP approximation of the detected process Ξ^-D is

$$f^{\Xi^+ D}(x) = m \Pr\{x \mid v_o\} = \sum_{r=1}^{m} \frac{l(y_r \mid x) f^{\Xi^- D}(x)}{\lambda(y_r)}, \quad x \in S^+ \tag{5.25}$$

Equivalently,

$$f^{\Xi^+ D}(x) = \sum_{r=1}^{m} \frac{l(y_r \mid x) P^D(x) f^{\Xi^-}(x)}{\lambda(y_r)} \tag{5.26}$$

where we have substituted (5.13) into (5.25) to obtain (5.26).

5.2.6 Correlation Losses in the PPP Approximation

The locations of points in a PPP are uncorrelated. Knowing the location of one point does not affect the distribution of the other points in the PPP. This is a consequence of the locations being obtained as i.i.d. draws from a given distribution. By contrast, in the Bayes posterior point process, point (target) locations are correlated. Knowing the location of one target tells us something about the location of other targets. In [7] it is shown the presence of a target at x in the posterior point process reduces the likelihood of finding another target nearby. This topic is discussed further in Section 5.7.1.

5.2.7 The iFilter

The undetected process $\Xi^- U$ is by definition unaffected by scan data, so its Bayes update is the predicted PPP whose intensity is given by (5.14). Because the detected and undetected processes are mutually independent, their Bayes updates are mutually independent as well. Thus, the intensity of the PPP approximation to the Bayes posterior process is the sum of the intensities of the Bayes updates of the detected and undetected processes; that is,

$$
\begin{aligned}
f^{\Xi^+}(x) &= f^{\Xi^+ U}(x) + f^{\Xi^+ D}(x) \\
&= (1 - P^D(x)) f^{\Xi^-}(x) + \sum_{r=1}^{m} \frac{l(y_r \mid x) P^D(x) f^{\Xi^-}(x)}{\lambda(y_r)} \\
&= f^{\Xi^-}(x) \left[1 - P^D(x) + \sum_{r=1}^{m} \frac{l(y_r \mid x) P^D(x)}{\lambda(y_r)} \right]
\end{aligned} \tag{5.27}
$$

For $m = 0$, the Bayes posterior process comprises only the predicted undetected process. Thus, for $m = 0$, the summation in (5.27) is omitted. Substituting the expression for λ given in (5.15) yields

$$f^{\Xi^+}(x) = f^{\Xi^-}(x)\left[1 - P^D(x) + \sum_{r=1}^{m} \frac{l(y_r \mid x) P^D(x)}{\lambda_\phi(y_r) + \int_S l(y_r \mid x) P^D(x) f^{\Xi^-}(x) \, dx}\right] \quad (5.28)$$

This is the intensity filter on the augmented space S^+.

Note that although all possible measurement-to-target assignments were used in the derivation, the iFilter recursion requires neither the enumeration of these assignments nor the computation of their probabilities. In contrast, such calculations are necessary in MHT methods.

5.2.8 Transformations of PPPs Are PPPs

In this section we provide proofs showing that a PPP subject to a Markov transformation or a thinning is a PPP. The derivations are given for the spaces S and T, but they also hold for augmented spaces such as S^+.

5.2.8.1 A Markov Transformation of a PPP Is a PPP

A PPP whose points undergo a Markovian transition is still a PPP. Let Ξ be a PPP on S, and let Ψ be a transition function on S, so that the probability that a point $x \in S$ transitions to $y \in T$ is $\Psi(y \mid x)$. If $\xi = (n, \{x_1, \ldots, x_n\})$ is a realization of the PPP Ξ with intensity $\lambda(x)$, the realization after the Markov transition is $\eta \equiv (n, \{y_1, \ldots, y_n\}) \in \mathcal{E}(T)$, where y_j is a realization of the pdf $\Psi(\cdot \mid x_j), j = 1, \ldots, n$. We now show that the transformed process, denoted $\Psi(\Xi)$, is a PPP on T with intensity

$$\nu(y) = \int_S \Psi(y \mid x) \lambda(x) \, dx \quad \text{for } y \in T$$

Let $\eta_o = (n, (y_1, \ldots, y_n))$ be the ordered realization of η defined above, and let

$$\mu_\lambda = \int_S \lambda(s) \, ds$$

Observe that

$$\mu_\nu \equiv \int_T \nu(y) \, dy = \int_T \int_S \Psi(y \mid x) \lambda(x) \, dx \, dy$$

$$= \int_S \int_T \Psi(y \mid x) \, dy \, \lambda(x) \, dx = \int_S \lambda(x) \, dx = \mu_\lambda$$

Then

$$\Pr\left\{\eta_o = (n,(y_1,\ldots,y_n))\right\}$$

$$= \int_S \cdots \int_S \prod_{j=1}^n \Psi(y_j \mid x_j) \Pr\left\{\xi_o = (n,(x_1,\ldots x_n))\right\} dx_1 \cdots dx_n$$

$$= \frac{e^{-\mu_\lambda}}{n!} \int_S \cdots \int_S \prod_{j=1}^n \Psi(y_j \mid x_j) \prod_{j=1}^n \lambda(x_j) \, dx_1 \cdots dx_n \qquad (5.29)$$

$$= \frac{e^{-\mu_\lambda}}{n!} \prod_{j=1}^n \int_S \Psi(y_j \mid x_j) \lambda(x_j) \, dx_j$$

$$= \frac{e^{-\mu_\lambda}}{n!} \prod_{j=1}^n \nu(y_j) = \frac{e^{-\mu_\nu}}{n!} \prod_{j=1}^n \nu(y_j)$$

Thus $\Psi(\Xi)$ is a PPP on T with intensity ν. This proof applies to the Markov motion update in Section 5.1.4 by setting $T = S$, $y = x'$, and $\Psi = q$. Similarly, we can show that the measurement process in Section 5.1.5 is a PPP by setting $\Psi = l$.

5.2.8.2 A Bernoulli Thinned PPP Is a PPP

Let Ξ be a point process on S with intensity λ. For every $x \in S$, let $1 - \alpha(x)$, $0 \le \alpha(x) \le 1$, be the probability that the point x is removed from any realization that contains it. For the realization $\xi = (n, \{x_1, \ldots, x_n\})$, the point x_j is retained with probability $\alpha(x_j)$ and dropped with probability $1 - \alpha(x_j)$. Thinning of the points is mutually independent. The thinned realization is $\xi_\alpha = (m, \{x_1', \ldots, x_m'\})$, where $m \le n$ is the number of points that survive the thinning. Let Ξ_α be the process that results from this thinning and define

$$\lambda_\alpha(x) = \alpha(x)\lambda(x) \quad \text{for } x \in S$$

We will show that Ξ_α is a PPP with intensity λ_a.

Let

$$\mu = \int_S \lambda(x)\,dx \quad \text{and} \quad \mu(\alpha) = \int_S \lambda_\alpha(x)\,dx$$

The probability that a point x is in a realization of Ξ is $\lambda(x)/\mu$. The probability that x is in Ξ and that it is also retained in Ξ_a is $\alpha(x)\lambda(x)/\mu$. Thus, the probability that a point in Ξ is retained in Ξ_a is

$$\frac{1}{\mu}\int_S \alpha(x)\lambda(x)dx = \frac{\mu(\alpha)}{\mu}$$

Let N and M be the random variables giving the number of points in the realizations of Ξ and Ξ_α respectively. For $n \geq m$,

$$\Pr\{M = m \mid N = n\} = \frac{n!}{m!(n-m)!}\left[\frac{\mu(\alpha)}{\mu}\right]^m\left[1 - \frac{\mu(\alpha)}{\mu}\right]^{n-m}$$

and

$$\Pr\left\{\xi_\alpha = \left(m,\{x_1',\ldots,x_m'\}\right) \mid M = m \text{ and } N = n\right\}$$

$$= \Pr\left\{\xi_\alpha = \left(m,\{x_1',\ldots,x_m'\}\right) \mid M = m\right\}$$

$$= m!\prod_{j=1}^m \frac{\alpha(x_j')\lambda(x_j')}{\mu(\alpha)}$$

Thus

$$\Pr\left\{\xi_\alpha = \left(m,\{x_1',\ldots,x_m'\}\right) \mid N = n\right\}$$

$$= \Pr\left\{\xi_\alpha = \left(m,\{x_1',\ldots,x_m'\}\right) \mid M = m \text{ and } N = n\right\}\Pr\{M = m \mid N = n\}$$

$$= m!\left(\prod_{j=1}^m \frac{\alpha(x_j')\lambda(x_j')}{\mu(\alpha)}\right)\frac{n!}{m!(n-m)!}\left[\frac{\mu(\alpha)}{\mu}\right]^m\left[1 - \frac{\mu(\alpha)}{\mu}\right]^{n-m}$$

Since $\Pr\{N = n\} = \mu^n e^{-\mu}/n!$,

$$\Pr\left\{\xi_\alpha = \left(m,\{x_1',\ldots,x_m'\}\right)\right\}$$

$$= m!\left(\prod_{j=1}^m \frac{\alpha(x_j')\lambda(x_j')}{\mu(\alpha)}\right)\sum_{n=m}^\infty \frac{n!}{m!(n-m)!}\left[\frac{\mu(\alpha)}{\mu}\right]^m\left[1 - \frac{\mu(\alpha)}{\mu}\right]^{n-m}\frac{\mu^n e^{-\mu}}{n!}$$

$$= \left(\prod_{j=1}^m \frac{\alpha(x_j')\lambda(x_j')}{\mu(\alpha)}\right)\mu(\alpha)^m e^{-\mu}\sum_{n=m}^\infty \frac{1}{(n-m)!}\left[1 - \frac{\mu(\alpha)}{\mu}\right]^{n-m}\mu^{n-m}$$

$$= \left(\prod_{j=1}^m \alpha(x_j')\lambda(x_j')\right)e^{-\mu}\sum_{n=m}^\infty \frac{1}{(n-m)!}(\mu - \mu(\alpha))^{n-m}$$

$$= \left(\prod_{j=1}^m \alpha(x_j')\lambda(x_j')\right)e^{-\mu(\alpha)}$$

which is the pdf for a PPP with intensity function $\lambda_\alpha(s) = \alpha(s)\lambda(s)$ for $s \in S$.

5.3 PHD FILTER

The PHD filter was derived in [2, 3] on the nonaugmented space S. However, it can also be derived from the iFilter (5.28) by making a judicious choice of the target transition function q and restricting the output of the filter to S. The measurement space T is the same in both filters.

The PHD filter uses a specified new target birth process with intensity $B(x)$ and measurement clutter process with intensity $\lambda_c(y)$. These processes are assumed to be PPPs that are mutually independent of, respectively, the existing target PPP and the target-originated measurement PPP. The PHD filter assumes we are given a target transition function $\tilde{q}(x \mid x')$ for all $x, x' \in S$.

Our goal is to define an iFilter transition function q on S^+ so that its information-updated intensity, when restricted to S, is identical to the PHD filter update using the transition function $\tilde{q}$ and the specified new target and clutter intensities. We begin by defining $q(x \mid x') \equiv \tilde{q}(x \mid x')$ for all $x, x' \in S$. Thus,

$$\int_{S^+} q(x \mid x')\,dx = q(\phi \mid x') + \int_S \tilde{q}(x \mid x')\,dx = q(\phi \mid x') + 1 \qquad (5.30)$$

For q to be a transition function on S, we must choose $q(\phi \mid x') = 0$, $x' \in S$.

The iFilter predicted target process Ξ^- is a PPP whose intensity is given by (5.12). Expanding the integral using the analog of (5.10) gives, for $x \in S^+$,

$$f^{\Xi^-}(x) = q(x \mid \phi)(1 - d(\phi)) f^{\Xi}(\phi) + \int_S \tilde{q}(x \mid x')(1 - d(x')) f^{\Xi}(x')\,dx' \qquad (5.31)$$

To match the predicted intensity of the iFilter to the predicted intensity of the PHD filter on S, we restrict (5.31) to $x \in S$ and define

$$q(x \mid \phi)(1 - d(\phi)) f^{\Xi}(\phi) = B(x) \qquad (5.32)$$

It follows that $q(x \mid \phi) \propto B(x)$. Integrating both sides of (5.32) over S gives

$$(1 - q(\phi \mid \phi))(1 - d(\phi)) f^{\Xi}(\phi) = \int_S B(x)\,dx \qquad (5.33)$$

This equation is a constraint on the choice of $q(\phi \mid \phi)$, $d(\phi)$, and $f^{\Xi}(\phi)$.

The intensity function of the predicted measurement process in (5.15) and (5.16) is identical to that of the PHD filter provided we define the iFilter clutter intensity to be

$$l(y \mid \phi) P^D(\phi) f^{\Xi^-}(\phi) = \lambda_c(y) \qquad (5.34)$$

Because $q(\phi \mid x') = 0$, $x' \in S$, the predicted intensity (5.31) for $x = \phi$ simplifies to $f^{\Xi^-}(\phi) = q(\phi \mid \phi)(1 - d(\phi)) f^{\Xi}(\phi)$. Substituting this into (5.34) and using (5.33) gives

$$l(y \mid \phi) P^D(\phi) \left(\frac{q(\phi \mid \phi)}{1 - q(\phi \mid \phi)} \right) \int_S B(x) \, dx = \lambda_c(y) \tag{5.35}$$

It follows that $l(y \mid \phi) \propto \lambda_c(y)$. Integrating both sides of (5.35) gives

$$P^D(\phi) \left(\frac{q(\phi \mid \phi)}{1 - q(\phi \mid \phi)} \right) \int_S B(x) \, dx = \int_T \lambda_c(y) \, dy \tag{5.36}$$

Let $P^D(\phi) > 0$ and $d(\phi) < 1$ be given. Assume that the expected numbers of new targets and clutter are nonzero. Then (5.36) has a unique solution, denoted $q^*(\phi \mid \phi)$. The solution satisfies $0 < q^*(\phi \mid \phi) < 1$. Substituting $q^*(\phi \mid \phi)$ into (5.33) determines the product $(1 - d(\phi)) f^{\Xi}(\phi)$ and, since $d(\phi) < 1$, this uniquely determines $f^{\Xi}(\phi)$.

With these definitions, the predicted target intensity of the iFilter for $x \in S$ is

$$f^{\Xi^-}(x) = B(x) + \int_S \tilde{q}(x \mid x')(1 - d(x')) f^{\Xi}(x') \, dx' \tag{5.37}$$

The intensity of the predicted measurement process for $y \in T$ is

$$\lambda(y) = \lambda_c(y) + \int_S l(y \mid x) P^D(x) f^{\Xi^-}(x) \, dx \tag{5.38}$$

and the intensity of the Bayes posterior point process for $x \in S$ is

$$f^{\Xi^+}(x) = f^{\Xi^-}(x) \left[1 - P^D(x) + \sum_{r=1}^m \frac{l(y_r \mid x) P^D(x)}{\lambda_c(y_r) + \int_S l(y_r \mid x) P^D(x) f^{\Xi^-}(x) \, dx} \right] \tag{5.39}$$

The iFilter expressions (5.37)–(5.39) are identical to the PHD filter.

Intuitive Interpretation of the Recursion

The filter update (5.39) can be interpreted intuitively as follows. The probability that the data point y_r originates from a target whose state is in the infinitesimal $(x, x + dx)$ with volume $|dx|$ is the (Bayesian) ratio

$$\frac{l(y_r \mid x) P^D(x) f^{\equiv^-}(x) \mid dx \mid}{\lambda_c(y_r) + \int_S l(y_r \mid x) P^D(x) f^{\equiv^-}(x) \, dx}$$

Because there is at most one target per measurement and the data are conditionally independent, the sum from $r = 1$ to m is the expected number of targets in $(x, x + dx)$ that generate measurements. The expected number of targets that do not generate measurements is

$$\left(1 - P^D(x)\right) f^{\equiv^-}(x) \mid dx \mid$$

The sum is the expected total number of targets in the infinitesimal. Assuming the posterior process is a PPP, the sum equals $f^{\equiv^+}(x) \mid dx \mid$. Dividing both sides of this equation by $\mid dx \mid$ gives the updated intensity per unit state space (5.39).

5.4 PGF APPROACH TO THE iFILTER

This section studies a seemingly very special class of multitarget tracking problems. In these problems, target state space is gridded and the measurement data are histogram counts (i.e., nonnegative integers in each grid cell). These problems are useful in applications in which data are aggregated, or binned, into histogram cells (for example, geospatial temporal data may only be available on discrete lattices [8, Chap. 4]). Moreover, the iFilter and PHD filter can be recovered in the "small cell limit." In this limit the cells are ultimately so small that the histogram counts are either zero or one, and cells with a count of one correspond to the points of a sensor scan.

One advantage of the discrete approach to the continuous limit is that the classical tools of PGFs for multivariate discrete random variables can be used. Many engineers and signal processors know the PGF by the name "z-transform." PGFs enable us not only to avoid less familiar mathematical methods needed for working in the continuous domain directly, but also to see clearly into the nature of the target and measurement modeling.

Wilf [9] wrote that, "A generating function is a clothesline on which we hang up a sequence of numbers for display." In target tracking applications, these numbers are probabilities that characterize the measurement-to-target assignments. Probabilities and expected values are recovered by differentiation, so it is reasonable to say that PGFs "encode" measurement-to-target assignment probabilities in their derivatives. As will be seen, these derivatives can become nontrivial—"decoding" the PGF becomes steadily more involved as the number and nature of feasible assignments becomes more numerous.

We review basic properties of PGFs and multivariate PGFs in the first subsection. Readers can skip this discussion entirely and refer to it as need arises. The ensuing subsections derive the joint PGF and corresponding iFilter for discrete target state and measurement histograms. The small cell limit of the iFilter is then shown to be the PHD filter.

5.4.1 Brief Review of PGFs

This section presents a brief review of the basic properties of PGFs.

5.4.1.1 Univariate PGFs

Let A be a discrete random variable with outcomes in $\mathbb{N} = \{0,1,2,...\}$. Let $p(a) = \Pr\{A = a\}$. The PGF of A is defined by the Maclaurin series

$$G(x) = \sum_{a=0}^{\infty} p(a) x^a$$

Because $G(1) = \sum_{a=0}^{\infty} p(a) = 1$, $G(x)$ is analytic in the complex x-plane in a domain that includes the closed disc $|x| \leq 1$. (Readers with signal processing backgrounds will notice that $G(z^{-1})$ is the z-transform of the probability sequence. The moment generating function and characteristic function of the sequence are $M(t) = G(e^t)$ and $\Phi(t) = G(e^{t\sqrt{-1}})$, respectively.) Since $G(x)$ is analytic at the origin,

$$G^{(a)}(0) \equiv \frac{d^a}{dx^a} G(x) \bigg|_{x=0} = a! \, p(a) \qquad (5.40)$$

so that $p(a) = G^{(n)}(0)/a!$. The mean of A is the derivative evaluated at $x = 1$; that is, $E[A] = G^{(1)}(1)$.

Example. Let A be Poisson-distributed with mean $\mu \geq 0$, so that (*cf.* (5.4)) $p(a) = e^{-\mu} \mu^a / a!$. From the definition of the PGF, it is easily seen that

$$G(x) = e^{-\mu + \mu x} \qquad (5.41)$$

is the PGF. Note that $E[A] = G'(1) = \mu$.

5.4.1.2 Multivariate PGFs and Bayes' Theorem

The case with four variables is illustrative of the general definitions and methods. The multivariate PGF of the vector (A,B,C,D) of random variables with outcomes $(a,b,c,d) \in \mathbb{N}^4$ is

$$G(x,y,z,w) = \sum_{a=0}^{\infty}\sum_{b=0}^{\infty}\sum_{c=0}^{\infty}\sum_{d=0}^{\infty} p(a,b,c,d)\, x^a y^b z^c w^d \qquad (5.42)$$

where $p(a,b,c,d) = \Pr\{A=a,\ B=b,\ C=c,\ D=d\}$ is the joint probability distribution function.

For example, histograms of i.i.d. data are modeled by the multinomial distribution. Let n denote the total number of samples in a four-cell histogram with pairwise disjoint cells. Assume the n samples are i.i.d. over the four cells with cell probabilities p_1, p_2, p_3, p_4, where $p_i \geq 0$ and $p_1 + p_2 + p_3 + p_4 = 1$. Let (A,B,C,D) denote the random vector of the number of samples in the four cells. The probability of a histogram with counts $(A,B,C,D) = (a,b,c,d)$ is

$$p(a,b,c,d) = \frac{n!}{a!b!c!d!}\, p_1^a p_2^b p_3^c p_4^d \qquad (5.43)$$

where $0 \leq a,b,c,d \leq n$ and $a+b+c+d = n$. The PGF of histograms with a specified number n of samples can be found be direct calculation. Only a finite number of terms in (5.42) are nonzero. Substituting (5.43) into (5.42) gives

$$G(x,y,z,w) = \sum_{\substack{a,b,c,d=0 \\ a+b+c+d=n}}^{n} \frac{n!}{a!b!c!d!}\, p_1^a p_2^b p_3^c p_4^d\, x^a y^b z^c w^d$$

$$= (p_1 x + p_2 y + p_3 z + p_4 w)^n \qquad (5.44)$$

This PGF extends to any number of histogram cells.

The series in (5.42) converges when the variables x,y,z,w are each in the closed unit disc. Note that the PGF of A is $G(x,1,1,1)$. Differentiating (5.42) term by term gives

$$p(a,b,c,d) = \frac{1}{a!b!c!d!}\, G^{(a,b,c,d)}(0,0,0,0)$$

$$\equiv \frac{\partial^{a+b+c+d}}{\partial x^a\, \partial y^b\, \partial z^c\, \partial w^d} G(x,y,z,w)\Bigg|_{x=y=z=w=0} \qquad (5.45)$$

An important fact is that PGF of the sum of mutually independent multivariate random variables (of comparable dimension, which is four in the example) is the product of their PGFs. The sum will sometimes be called a superposition when the random variables are counting variables.

Ratios of derivatives of the joint PGF produce PGFs of conditional distributions and in this sense encode Bayes' theorem. Conditioning on, say $B = b$, gives the trivariate random variable, $(A,C,D) \mid B = b$. Substituting conditional probabilities into the definition of the PGF gives

$$
\begin{aligned}
G(x,z,w \mid b) &\equiv \sum_{a=0}^{\infty} \sum_{c=0}^{\infty} \sum_{d=0}^{\infty} p(a,c,d \mid b) \, x^a z^c w^d \\
&= \sum_{a=0}^{\infty} \sum_{c=0}^{\infty} \sum_{d=0}^{\infty} \frac{p(a,b,c,d)}{p(b)} x^a z^c w^d \\
&= \frac{G^{(0,b,0,0)}(x,0,z,w)}{G^{(0,b,0,0)}(1,0,1,1)}
\end{aligned}
\tag{5.46}
$$

where $G^{(0,b,0,0)}(1,0,1,1) = b! \, p(b)$. Equation (5.46) is the PGF of the Bayes posterior distribution for $(A,C,D) \mid B = b$. Using (5.45) gives the conditional probabilities

$$
p(a,c,d \mid b) = \frac{1}{a! c! d!} \frac{G^{(a,b,c,d)}(x,y,z,w)}{G^{(0,b,0,0)}(1,0,1,1)} \Bigg|_{x=y=z=w=0}
\tag{5.47}
$$

These results extend to any number of variables.

The PGF of the conditional variable $A \mid B = b$ is found by substituting $z = w = 1$ into (5.46). The conditional mean is thus the derivative of (5.46) with respect to x evaluated at $x = z = w = 1$:

$$
E[A \mid B = b] = \frac{G^{(1,b,0,0)}(1,0,1,1)}{G^{(0,b,0,0)}(1,0,1,1)}
\tag{5.48}
$$

The conditional means of C and D are found in a similar manner.

5.4.1.3 PGFs of Sums

The sum $N = A + B + C + D$ is a random integer. Let $G^{\#}(s)$ denote its PGF. From (5.42), we have

$$G(s,s,s,s) = \sum_{a=0}^{\infty}\sum_{b=0}^{\infty}\sum_{c=0}^{\infty}\sum_{d=0}^{\infty} p(a,b,c,d)\, s^{a+b+c+d}$$

$$= \sum_{n=0}^{\infty} s^n \sum_{a+b+c+d=n} p(a,b,c,d) = \sum_{n=0}^{\infty} \Pr[n]s^n \equiv G^{\#}(s)$$

The result holds for conditional PGFs too; for example, the PGF of the sum $(A+C+D)\,|\,B=b$ is, from (5.46), $G(s,s,s\,|\,b) \equiv G^{\#}(s\,|\,b)$.

5.4.1.4 PGFs of Random Histograms

As in the multinomial example in Section 5.4.1.2, let (A,B,C,D) denote the random number of samples in a four-cell histogram with pairwise disjoint cells where samples are i.i.d. with cell probabilities p_1, p_2, p_3, p_4. What is different here is that the total number of histogram samples is a random integer N whose probability distribution is assumed known. As in Section 5.4.1.3, let $G^{\#}(s)$ denote its PGF. With a four-cell histogram $N = A+B+C+D$. The joint PGF of (A,B,C,D) is

$$G(x,y,z,w) = \sum_{a=0}^{\infty}\sum_{b=0}^{\infty}\sum_{c=0}^{\infty}\sum_{d=0}^{\infty} p(a,b,c,d)\, x^a y^b z^c w^d$$

$$= \sum_{n=0}^{\infty} \Pr\{N=n\} \sum_{\substack{a,b,c,d=0 \\ a+b+c+d=n}}^{n} p(a,b,c,d\,|\,n)\, x^a y^b z^c w^d \qquad (5.49)$$

where $p(a,b,c,d\,|\,n) = \Pr\{a,b,c,d|n\}$. The inner sum is the conditional histogram PGF conditioned on $N = n$ samples, which we denote by $G(x,y,z,w\,|\,n)$. It is recognized from (5.44):

$$G(x,y,z,w\,|\,n) = (p_1 x + p_2 y + p_3 z + p_4 w)^n \qquad (5.50)$$

Substituting (5.50) into (5.49) gives

$$G(x,y,z,w) = \sum_{n=0}^{\infty} \Pr\{N=n\}(p_1 x + p_2 y + p_3 z + p_4 w)^n \qquad (5.51)$$

$$= G^{\#}(p_1 x + p_2 y + p_3 z + p_4 w)$$

The result (5.51) plays a key role in deriving intensity filter recursions.

5.4.2 The iFilter on Finite Grids

Multitarget states and measurements are modeled using finite point processes that are defined on a finite grid of discrete state and measurement histogram cells. The joint PGF, or multivariate z-transform, is shown to be the composition of the PGF of the number of targets and a conditional measurement PGF.

Conditioned on measurement data, the domain of definition of the Bayes posterior process is the set of discrete target states. The PGF of the posterior is a ratio of ordinary derivatives of the joint PGF.

Several summary statistics of the Bayes posterior process are derived from the PGF. These include the mean and variance of the number of targets in each target state (histogram cell), as well as the mean and variance of the total number of targets.

In the tracking application, summary statistics are used to approximate the Bayesian posterior process and close the Bayesian recursion. Although all the summary statistics are of application interest, the intensity filter depends only on the intensity function. The intensity is the *sufficient* statistic for the discrete Poisson process that approximates the Bayes posterior process. The set of intensities, one for each discrete target state, comprises the output of the intensity filter.

5.4.3 Joint PGF of Gridded States and Histogram Data

This section develops the iFilter on discrete finite grids. We parallel the notation of Section 5.2.2 for the previous, predicted, and information-updated processes. Thus, we denote the PGFs of target state and total target number at the previous time step by $G(\cdot)$ and $G^{\#}(\cdot)$, and we denote the PGFs and intensities associated with the predicted target process by $G^{-}(\cdot)$, $G^{\#-}(\cdot)$, and $\lambda^{-}(\cdot)$. Finally, we denote the information updated target PGFs and intensities by $G^{+}(\cdot)$, $G^{\#+}(\cdot)$, and $\lambda^{+}(\cdot)$. These distinctions are not needed for the PGFs and intensities of the new target and measurement processes.

5.4.3.1 State Grid and Prior Target PGF

The state space S is assumed to be bounded. Partition it into a grid of $n \geq 1$ cells and denote the grid cells by $X_1, ..., X_n$. We are modeling the state space distribution of multiple target count, so let $N(X_i)$ denote the number of targets in cell X_i. Denote the total number of targets by $N = N(X_1) + \cdots + N(X_n)$.

Targets at the previous time scan are assumed to be located in grid cells determined by i.i.d. draws from the discrete distribution whose cell probabilities are denoted by $p(X_i)$, $i = 1, ..., n$. The multivariate PGF of the random counts $\left(N(X_1), ..., N(X_n)\right)$ at the previous time step is by definition (*cf.* (5.42)),

$$G(x_1,...,x_n) = \sum_{k_1=0}^{\infty} \cdots \sum_{k_n=0}^{\infty} \Pr\{k_1,...,k_n\} x_1^{k_1} \cdots x_n^{k_n} \qquad (5.52)$$

where $\Pr\{k_1,...,k_n\}$ is the probability of $(N(X_1)=k_1,...,N(X_n)=k_n)$. The target distribution assumptions are the same as were used in obtaining (5.49); thus, the PGF of the target count histogram is

$$G(x_1,...,x_n) = G^\#\left(\sum_{i=1}^{n} p(X_i)x_i\right) \qquad (5.53)$$

where, as mentioned above, $G^\#(\cdot)$ is the PGF of the prior distribution of N, the total number of targets in the state histogram at the previous time scan. This distribution can take a specified parametric form, or it can be specified explicitly by giving the probabilities $\Pr\{N=n\}$.

Assuming that the number of targets at the previous time scan is Poisson-distributed gives, from (5.41),

$$G^\#(s) = e^{-\lambda+\lambda s} \qquad (5.54)$$

where λ is the expected total number of targets. The expected number of targets in cell X_i at the previous time step is therefore

$$\lambda(X_i) = E[N(X_i)] \equiv \lambda p(X_i) \qquad (5.55)$$

The expected number λ and probabilities $p(X_i)$ are assumed known. After the initial step, they are specified via the filter recursion.

5.4.3.2 Predicted Target PGF

Realizations of the prior point process on the discrete space $X_1,...,X_n$ are subjected to independent thinning with survival probability $1-d(X_i)$, where $d(X_i)$ is the probability that a target in cell X_i dies at the previous time scan. Surviving targets may move, or propagate, to another target cell at the current time; they may also stay in the same cell. Let $p(X_i \mid X_{i'})$ be the probability that a surviving target in cell $X_{i'}$ at the previous time scan moves to target cell X_i at the current time. (This is the discrete space version of the target motion model $q(x \mid x')$ on the continuous space.) Finally, let a Poisson-distributed number of new targets be added, or superimposed, with the surviving targets after target motion. Let λ^B denote the mean number of new targets, and let $p^B(X_i)$ be the probability that a new target is birthed in cell X_i. Then

$$\lambda^{B}(X_{i}) \equiv \lambda^{B} p^{B}(X_{i}) \tag{5.56}$$

is the mean number of new targets in cell X_{i}.

Altogether these assumptions give the discrete analog to (5.39),

$$\lambda^{-}(X_{i}) = \lambda^{B}(X_{i}) + \sum_{j=1}^{n} p(X_{i} \mid X_{j})(1 - d(X_{j}))\lambda(X_{i}) \tag{5.57}$$

where $\lambda^{-}(X_{i})$ is the predicted expected number of targets in cell X_{i}. The properties of independent thinning (*cf.* (5.9)), target motion (*cf.* (5.7)), and superposition of independent PPPs hold for PPPs on discrete state spaces. Hence, the predicted target process at the current time is a discrete PPP whose cell means are given by (5.57).

From the assumptions, the predicted total number of targets is Poisson-distributed and has the form (5.54) but with parameter

$$\lambda^{-} = \sum_{i=1}^{n} \lambda^{-}(X_{i}) \tag{5.58}$$

The predicted target cell probabilities $p^{-}(X_{i})$ are defined via the equation

$$\lambda^{-}(X_{i}) = \lambda^{-} p^{-}(X_{i}) \tag{5.59}$$

Therefore, the PGF of the predicted target process is

$$G^{-}(x_{1},...,x_{n}) = G^{\#-}\left(\sum_{i=1}^{n} p^{-}(X_{i})x_{i}\right) \tag{5.60}$$

where under PPP assumptions $G^{\#-}(s) = \exp(-\lambda^{-} + \lambda^{-} s)$.

5.4.3.3 Conditional PGF of the Measurement Histogram

The measurement space is assumed to be bounded and partitioned into $m \geq 1$ grid cells. These cells are denoted by $Y_{1},...,Y_{m}$. Because of the gridding the sensor scan is not a list of point measurements, but a histogram of nonnegative integer counts J_{i} in the measurement cells. Intuitively, if the cells are sufficiently small, the usual scan of point measurements maps into a histogram whose entries are one or zero depending on whether or not the cell contains a point measurement. Let the total number of measurements be $J = J(Y_{1}) + \cdots + J(Y_{m})$.

Let $1 - P^{D}(X_{\ell})$ be the probability that a target in cell X_{ℓ} does not generate a measurement in any of the measurement cells $Y_{1},...,Y_{m}$. Let $p(Y_{i} \mid X_{\ell})$ be the conditional probability that given a target in cell X_ℓ generates a measurement that falls in cell Y_i. The multivariate PGF for one target in cell X_ℓ is

$$G_\ell(y_1,...,y_m) \equiv G_\ell(y_1,...,y_m \mid N(X_\ell)=1)$$
$$= 1 - P^D(X_\ell) + P^D(X_\ell)\sum_{i=1}^m p(Y_i \mid X_\ell)y_i \tag{5.61}$$

This follows from the general version of (5.42) and the fact that the sums are over only those indices whose sum is either one or zero. If there are $N(X_\ell) = k_\ell \geq 0$ targets in cell X_ℓ, they are assumed to generate measurements independently of one another. The measurement histogram from targets in X_ℓ is the superposition of these measurements. The PGF of the superposition of mutually independent processes is the product of their individual PGFs, and these PGFs are identical by assumption, so

$$G_\ell(y_1,...,y_m \mid N(X_\ell)=k_\ell) = \left(G_\ell(y_1,...,y_m)\right)^{k_\ell} \tag{5.62}$$

Targets in different cells are assumed to generate mutually independent measurement histograms, so we have the conditional PGF of all measurements:

$$G(y_1,...,y_m \mid k_1,...,k_n) = \prod_{\ell=1}^n \left(G_\ell(y_1,...,y_m)\right)^{k_\ell} \tag{5.63}$$

As will be seen, it is important that the factors in (5.63) occur with powers k_ℓ equal to the numbers of targets in the histogram cells.

The measurement process is the superposition of target-originated and clutter-originated measurement processes. These processes are assumed mutually independent. Therefore, the PGF of the total measurement process is the product

$$G(y_1,...,y_m \mid k_1,...,k_n) = G_c(y_1,...,y_m)G(y_1,...,y_m \mid k_1,...,k_n)$$
$$= G_c(y_1,...,y_m)\prod_{\ell=1}^n \left(G_\ell(y_1,...,y_m)\right)^{k_\ell} \tag{5.64}$$

where $G_c(\cdot)$ is the joint PGF of the clutter measurement counts.

Clutter is modeled as a PPP on the discrete space $Y_1,...,Y_m$ with assumed cell intensities $\lambda_c(Y_i)$, $i=1,...,m$. Let

$$\lambda_c = \sum_{i=1}^m \lambda_c(Y_i) \tag{5.65}$$

denote the mean total number of clutter measurements. The probability $p(Y_i)$ of a clutter measurement in cell Y_i is thus defined implicitly via the equation

$$\lambda_c(Y_i) = \lambda_c p(Y_i) \tag{5.66}$$

Clutter is a PPP, so its PGF is given by

$$G_c(y_1,...,y_m) = G_c^{\#}\left(\sum_{i=1}^{m} p(Y_i)y_i\right) \tag{5.67}$$

where $G_c^{\#}(s) = \exp(-\lambda_c + \lambda_c s)$ is the PGF of the total number of clutter measurements and has the form given by (5.54).

5.4.3.4 Joint Target-Measurement Histogram PGF

Let $(j_1,...,j_m)$ denote a realization of $(J(Y_1),...,J(Y_m))$. The joint multivariate PGF of target-measurement cell counts is, from the pattern exemplified by (5.42),

$$
\begin{aligned}
&G(y_1,...,y_m,x_1,...,x_n) \\
&= \sum_{j_1=0}^{\infty} \cdots \sum_{j_m=0}^{\infty} \sum_{k_1=0}^{\infty} \cdots \sum_{k_n=0}^{\infty} \Pr\{j_1,...,j_m,k_1,...,k_n\} y_1^{j_1} \cdots y_m^{j_m} x_1^{k_1} \cdots x_n^{k_n}
\end{aligned}
\tag{5.68}
$$

From Bayes' theorem,

$$\Pr\{j_1,...,j_m,k_1,...,k_n\} = \Pr\{k_1,...,k_n\}\Pr\{j_1,...,j_m \,|\, k_1,...,k_n\}$$

Substituting this factorization into (5.68) yields

$$
\begin{aligned}
&G(y_1,...,y_m,x_1,...,x_n) \\
&= \sum_{k_1=0}^{\infty} \cdots \sum_{k_n=0}^{\infty} \Pr\{k_1,...,k_n\} G(y_1,...,y_m \,|\, k_1,...,k_n) x_1^{k_1} \cdots x_n^{k_n}
\end{aligned}
\tag{5.69}
$$

where the conditional PGF is defined by

$$
\begin{aligned}
&G(y_1,...,y_m \,|\, k_1,...,k_n) \\
&= \sum_{j_1=0}^{\infty} \cdots \sum_{j_m=0}^{\infty} \Pr\{j_1,...,j_m \,|\, k_1,...,k_n\} y_1^{j_1} \cdots y_m^{j_m}
\end{aligned}
\tag{5.70}
$$

It is unnecessary to perform the summations in (5.70) since we already know that the conditional PGF is given by (5.64). Substituting (5.64) into (5.69) gives

$$G(y_1,...,y_m,x_1,...,x_n)$$
$$= G_c(y_1,...,y_m)\sum_{k_1=0}^{\infty}\cdots\sum_{k_n=0}^{\infty}\Pr\{k_1,...,k_n\}\prod_{\ell=1}^{n}(x_\ell\,G_\ell(y_1,...,y_m))^{k_\ell} \quad (5.71)$$

The multiple summation on the right-hand side is immediately recognized as the definition of the PGF (5.52) evaluated at the arguments $x_\ell G_\ell(y_1,...,y_m)$, $\ell=1,...,n$. This PGF is the histogram PGF (5.53). The joint PGF of the target-measurement cell counts is

$$G(y_1,...,y_m,x_1,...,x_n)= G_c^{\#}\left(\sum_{i=1}^{m}p(Y_i)y_i\right)G^{\#}\left(\sum_{i=1}^{n}p(X_i)x_i\,G_i(y_1,...,y_m)\right) (5.72)$$

The composition of PGFs is a prominent feature of (5.72). (It is also one of the hallmarks of branching processes. See the discussions in Sections 5.6 and 5.7.)

5.4.3.5 Posterior Process

The PGF (5.72) is fully specified once the PGFs $G^{\#}(\cdot)$, $\{G_i(\cdot), i=1,...,m\}$, and $G_c^{\#}(\cdot)$ are known. Under PPP assumptions the PGFs of the numbers of clutter and predicted targets are exponentials, while the PGFs of the sensor measurement process are defined by (5.61). Substituting these expressions gives, explicitly,

$$G(y_1,...,y_m,x_1,...,x_n)= \exp\left(-\lambda_c+\sum_{i=1}^{m}\lambda_c(Y_i)y_i\right)$$
$$\times\exp\left(-\lambda^-+\sum_{\ell=1}^{n}x_\ell\,\lambda^-(X_\ell)\left(1-P^D(X_\ell)+P^D(X_\ell)\sum_{i=1}^{m}p(Y_i\,|\,X_\ell)y_i\right)\right) \quad (5.73)$$

The posterior distribution on the integer counts $(k_1,...,k_n)$ in target space cells $X_1,...,X_n$ is determined by Bayes' theorem once the measurements are given.

The measurement histogram consists of the integer counts $(j_1,...,j_m)$ in the measurement space cells $Y_1,...,Y_m$. From the general version of (5.46), the PGF of the distribution of the information updated process is the ratio of derivatives

$$G^{+}(x_1,...,x_n\,|\,j_1,...,j_m)=\frac{G^{(j_1,...,j_m,0,...,0)}(0,...,0,x_1,...,x_n)}{G^{(j_1,...,j_m,0,...,0)}(0,...,0,1,...,1)} \quad (5.74)$$

The high order derivatives of (5.73) with respect to the variables y_i are readily evaluated because $\log G$ is linear in these variables. Thus,

$$G^{(j_1,...,j_m,0,...,0)}(0,...,0,x_1,...,x_n)= \tilde{G}(x_1,...,x_n)A(x_1,...,x_n) \quad (5.75)$$

where

$$\tilde{G}(x_1,...,x_n) = \exp\left(-\lambda_c - \lambda^- + \sum\nolimits_{\ell=1}^{n} x_\ell \lambda^- (X_\ell)(1 - P^D(X_\ell))\right)$$

$$A(x_1,...,x_n) = \prod_{i=1}^{m}\left(\lambda_c(Y_i) + \sum\nolimits_{\ell=1}^{n} x_\ell \lambda^- (X_\ell) P^D(X_\ell) p(Y_i \mid X_\ell)\right)^{j_i} \qquad (5.76)$$

Dividing as prescribed by (5.74) gives the conditional PGF of the information process as a product of PGFs:

$$G^+(x_1,...,x_n \mid j_1,...,j_m)$$

$$= \prod_{\ell=1}^{n} \exp\left(-\lambda^-(X_\ell)(1 - P^D(X_\ell)) + x_\ell \lambda^- (X_\ell)(1 - P^D(X_\ell))\right) \qquad (5.77)$$

$$\times \prod_{i=1}^{m}\left(\frac{\lambda_c(Y_i) + \sum_{\ell=1}^{n} x_\ell \lambda^- (X_\ell) P^D(X_\ell) p(Y_i \mid X_\ell)}{\lambda_c(Y_i) + \sum_{\ell=1}^{n} \lambda^- (X_\ell) P^D(X_\ell) p(Y_i \mid X_\ell)}\right)^{j_i}$$

The PGF shows that the Bayes posterior process has an interesting statistical interpretation. As can be seen from the product form, the posterior is the superposition of $n + m$ mutually independent point processes. The first n factors are the PGFs of PPPs, one for each target cell X_ℓ. The intensity of the PPP in cell X_ℓ is $\lambda^-(X_\ell)(1 - P^D(X_\ell))$. The second set of m factors corresponds to multinomial trials. The multinomial trial for measurement cell Y_i has $n+1$ outcomes and is repeated as many times as there are measurements in cell Y_i, namely, j_i. The outcomes of a multinomial trial are the number of ways a measurement can be assigned either to a target (there are n target cells X_i) or to clutter.

5.4.3.6 Summary Statistics: The iFilter and the Variances

The PGF (5.77) provides more insight into the structure of the Bayes posterior point process than does the direct method presented in Section 5.2. That method had the advantage of showing clearly the combinatorial nature of the iFilter, an aspect that is almost invisible in the PGF approach.

Several important facts are clear by inspection of the PGF (5.77). One is that the Bayes posterior process cannot be a PPP because the PGF $G^+(\cdot)$ does not have the right form. Consequently, the Bayes recursion is not closed and some kind of approximation is necessary in order to close it.

Moreover, since the PGF (5.77) is not a product of the form $\prod_{i=1}^{n} g_i(x_i)$ for any choice of univariate functions $g_i(\cdot)$, the numbers $N^+(X_i)$ cannot be mutually independent, where $N^+(X_i)$ denotes the number of targets in the Bayes posterior process in state cell X_i. In sharp contrast, the numbers of targets in the prior process, namely $N(X_i), i = 1,...,n$, are mutually independent. The cell-to-cell correlation is a consequence of the sensor observation process.

The iFilter approximates the Bayes posterior distribution with that of a PPP. To do this, we match the mean of the PPP to the mean of the Bayes posterior. The mean of the posterior distribution in each of the state cells X_i can be evaluated by differentiating the PGF (5.77) as prescribed via the example (5.48). This procedure is straightforward, but an easier method can be used in this case because of the statistical interpretation of the PGF mentioned in the previous subsection.

The marginal distribution of $N^+(X_i)$ is the sum of the posterior distribution of $(N^+(X_1),..., N^+(X_n))$ over the number of targets in every cell except those in X_i. The PGF of $N^+(X_i)$ is the PGF (5.77) evaluated at $x_\ell = 1, \ell \neq i$:

$$G^+\left(x_i \mid j_1,..., j_m\right) \equiv G^+\left(1,...,1, x_i ,1,...,1 \mid j_1,..., j_m\right) \qquad (5.78)$$

It follows from the product form of (5.77) that the PGF of $N^+(X_i)$ is a product of $m+1$ terms in x_i. The distribution of $N^+(X_i)$ is therefore the sum of $m+1$ mutually independent random variables. One is Poisson-distributed, and its mean is its parameter, $\lambda^-(X_i)(1 - P^D(X_i))$. The other variables are binomial trials, one for measurement cell Y_r. Each trial is repeated j_r times, so its mean is the coefficient of x_i multiplied by j_r. The mean of the sum is the sum of the means, so the mean of the posterior distribution of target count in cell X_i is

$$\lambda^+(X_i) \equiv E\left[N^+\left(X_i\right)\right]$$
$$= \lambda^-(X_i)\left[1 - P^D(X_i) + \sum_{r=1}^{m} j_r \frac{P^D(X_i)p(Y_r \mid X_i)}{\lambda_c(Y_r) + \sum_{\ell=1}^{n} \lambda^-(X_\ell)P^D(X_\ell)p(Y_r \mid X_\ell)}\right] \qquad (5.79)$$

The intensity (5.79) is the output of the iFilter for histogram cell $X_i, \ i = 1,...,n$.

Denote the total number of targets in the posterior distribution by $N^+ = \sum_{i=1}^{n} N^+(X_i)$. Although the numbers $N^+(X_i)$ are correlated, the mean of the sum is still the sum of the means. Thus, the mean total number of targets in the posterior process is

$$\lambda^+ \equiv E\left[N^+\right] = \sum_{i=1}^{n} E\left[N^+(X_i)\right] = \sum_{i=1}^{n} \lambda^+(X_i) \qquad (5.80)$$

Let $p^+(X_i)$ denote the posterior probability that a target is in cell X_i. It is determined implicitly via the equation

$$\lambda^+(X_i) = \lambda^+ p^+(X_i) \qquad (5.81)$$

Variance of Number of Targets per Cell

The variance of $N^+(X_i)$ also follows from its marginal PGF (5.78). The factors correspond to conditionally independent random processes, so their variances add. The variance of the Poisson process is its mean $\lambda^-(X_i)(1-P^D(X_i))$. The remaining m distributions are binomial with j_r trials, $r=1,...,m$, and so have variance $j_r p_r(1-p_r)$, where p_r is the coefficient of x_i. Therefore,

$$\mathrm{var}\left[N^+(X_i)\right] = \lambda^-(X_i)\left(1 - P^D(X_i)\right)$$
$$+ \sum_{r=1}^{m} j_r \frac{\lambda^-(X_i)P^D(X_i)p(Y_r \mid X_i)\left[\lambda_c(Y_r) + \sum_{\ell=1,\ell\neq i}^{n}\lambda^-(X_\ell)P^D(X_\ell)p(Y_r \mid X_\ell)\right]}{\left[\lambda_c(Y_r) + \sum_{\ell=1}^{n}\lambda^-(X_\ell)P^D(X_\ell)p(Y_r \mid X_\ell)\right]^2} \qquad (5.82)$$

Comparing the result to (5.79) shows that the variance of the number of targets in a cell in the Bayes posterior is always smaller than the variance (and mean) of this quantity in the approximating PPP.

The variance of the total number of targets N^+ cannot be found simply by adding the variances $\mathrm{var}[N^+(X_i)]$. The reason is that the numbers of targets $N^+(X_i)$ in different target cells are correlated, and the variance of a sum is the sum of the variances only if the variables are mutually independent. Fortunately, there is an easier method to find the variance of N^+. (For more discussion of correlation between cells, see comments in Section 5.7.1.)

Variance of the Total Number of Targets

The PGF of N^+ is found by generalizing the result of Section 5.4.1.3. Thus, setting $x_1 = \cdots = x_n = x$ in (5.77) gives

$$G_{N^+}^{\#}\left(x \mid j_1,...,j_m\right) = e^{-\lambda^U + \lambda^U x} \prod_{i=1}^{m}\left(\overline{q}_i + \overline{p}_i\, x\right)^{j_i} \qquad (5.83)$$

where the mean number of (predicted) undetected targets is

$$\lambda^U = \sum_{\ell=1}^{n} \lambda^-(X_\ell)\left(1 - P^D(X_\ell)\right)$$

and

$$\bar{p}_i = 1 - \bar{q}_i = \frac{\sum_{\ell=1}^n \lambda^-(X_\ell) P^D(X_\ell) p(Y_i \mid X_\ell)}{\lambda_c(Y_i) + \sum_{\ell=1}^n \lambda^-(X_\ell) P^D(X_\ell) p(Y_i \mid X_\ell)}$$

is the conditional probability that a measurement in cell Y_i originates from a target. The PGF is therefore the convolution of a Poisson-distributed number with mean λ^U and m Bernoulli trials, each repeated j_i times with probability of target present in the ith trial given by $\bar{p}_i$.

The mean of N^+ is the sum of the means of these distributions, thus confirming the result (5.80). The random integers are conditionally independent, as seen from the product form of (5.83). The variance of N^+ is therefore the sum of the variances of $m+1$ distributions:

$$\text{var}\left[N^+\right] = \lambda^U + \sum_{i=1}^n j_i \left(1 - \bar{p}_i\right) \bar{p}_i \tag{5.84}$$

5.4.4 Small Cell Size Limits

The discussion above does not require that histogram cells be a partition of some underlying Euclidean state and measurement space—the cells can in fact correspond to a very general discrete set of n elements. In these more general cases, the notion of a small grid limit may or may not be meaningful.

However, when the histograms are gridded partitions of bounded Euclidean spaces, the limit does exist under mild assumptions. For example, as the number of cells goes to infinity and the sizes of the target and measurement cells all go to zero, the discrete sums become Riemann sums that, in the small cell limit, become integrals—provided the underlying functions are sufficiently well-behaved (e.g., differentiable). A more rigorous analysis can be found in Moyal [5].

The iFilter has such a limit. To see this, we need only make the identification that histogram cells correspond to some well specified location in their interiors:

$$X_i \to x \in S$$
$$Y_i \to y \in T$$

We begin by taking the limit of the predicted process (5.57). Some care is required because there are infinitesimals dx at the previous time and at the current time. Intensities and likelihood functions defined on the cells are functions of the cell locations and are multiplied by infinitesimals corresponding to the cell size. From (5.55) and (5.56),

$$\lambda(X_i) = \lambda p(X_i) \to f^{\equiv}(x)\,dx$$
$$\lambda^B(X_i) = \lambda^B p^B(X_i) \to B(x)\,dx \tag{5.85}$$
$$p(X_i \mid X_j) \to q(x \mid x')\,dx$$

where $f^{\equiv}(x)$ is the limiting PPP intensity function of the target process at the previous time step and $q(x \mid x')$ is the target motion model. The probability of target death is a thinning probability, so that $d(X_i) \to d(x)$. The left-hand side and right-hand side of the predicted process (5.57) are multiplied by dx. Dividing by dx and taking the limit gives the predicted process intensity

$$\frac{\lambda^-(X_i)}{dx} \to B(x) + \int_S q(x \mid x')(1 - d(x'))f^{\equiv}(x')\,dx' = f^{\equiv^-}(x) \tag{5.86}$$

Similarly, from (5.81), (5.59), and (5.66), we have for the information update

$$\lambda^+(X_i) = \lambda^+ p^+(X_i) \to f^{\equiv^+}(x)\,dx$$
$$\lambda^-(X_i) = \lambda^- p^-(X_i) \to f^{\equiv^-}(x)\,dx$$
$$\lambda_c(Y_r) = \lambda_c p_c(Y_r) \to \lambda_c(y)\,dy$$
$$p(Y_r \mid X_i) \to l(y \mid x)\,dy$$

where the function $l(y \mid x)$ is the sensor likelihood function, and where $f^{\equiv^-}(x)$ and $f^{\equiv^+}(x)$ are the limiting intensity functions of the predicted and updated target processes, respectively. The probability of detection is a thinning probability and does not require multiplication by an infinitesimal, so that $P^D(X_r) \to P^D(x)$. With these identifications,

$$\lambda_c(Y_r) + \sum_{\ell=1}^{n} p(Y_r \mid X_\ell)P^D(X_\ell)\lambda^-(X_\ell)$$
$$\to \left(\lambda_c(y_r) + \int_S l(y_r \mid x)P^D(x)f^{\equiv^-}(x)\,dx \right) dy$$

and

$$\lambda^-(X_i)P^D(X_i)\,p(Y_r \mid X_i) \to l(y_r \mid x)P^D(x)f^{\equiv^-}(x)\,dx\,dy$$

The infinitesimal dy cancels out of the limiting ratios of these quantities in (5.79) and (5.82). Since

$$\lambda^+(X_i) \to f^{\Xi^+}(x)\, dx$$
$$\lambda^-(X_i)\left(1 - P^D(X_i)\right) \to \left(1 - P^D(x)\right)f^{\Xi^-}(x)\, dx$$

the infinitesimal dx also cancels because it appears on both sides of (5.79).

The total number of measurements $j = j_1 + \cdots + j_m$ is specified and held fixed in the small cell limit. It is assumed that j_ℓ becomes, for sufficiently small cells, either one or zero. This is equivalent to requiring that the limiting intensity have no Dirac delta function components. Such "impulsive" components are excluded because they can generate multiple measurements in the same location even in the limit. The measurement set therefore becomes the locations $(y_1, ..., y_{m'})$, where m' is the number of measurement cells having exactly one measurement.

Therefore, under these assumptions, in the small cell limit, the filter (5.79) becomes identical to (5.39) on the space S. The small cell limit affects the PGF also—it becomes the probability generating functional (PGFL) of the PHD filter. The PHD filter can be derived directly from the PGFL using the functional derivatives defined in the manner as in the classical calculus of variations, but with respect to a variation with respect to a Dirac delta function (more precisely, the limit of a sequence of variations that form a "test function" sequence for the Dirac delta function). These considerations would take us far afield and are not presented here.

Other Small Cell Limits

As mentioned above, it is not necessary that the histogram cells partition a Euclidean state space. The augmented space S^+ is one such example. In this case one of the discrete state space cells corresponds to ϕ and the remaining histogram cells partition the space S. Thus, the small cell limit is taken only over the cells that partition S. With this and other straightforward modifications, the resulting filter is identical to the intensity filter (5.28) on the augmented space S^+.

More generally, in some applications the underlying state space can be the direct sum of several discrete spaces. (For example, such models are used in [10] to model target and clutter spaces separately.) None of these spaces need be partitions of a Euclidean space; however, the small cell limit can be taken in those spaces that are Euclidean, leaving the other spaces unchanged. The resulting intensity filter would then be defined on the direct sum of discrete and Euclidean spaces, as the case may be, but the essential nature of the intensity filter is unchanged.

5.5 EXTENDED TARGET FILTERS

Extended targets are targets that produce more than one point measurement in a sensor scan. Such targets can be modeled in the PGF approach if the PGF of the number of measurements is known, and if the point measurements themselves can be modeled as i.i.d samples from a known spatial distribution.

This is accomplished by replacing the traditional point target model (5.61) with the assumed extended target model. Denote the PGF of the number of points an extended target can produce in the sensor by

$$G_{\text{extended}}^{\#}(s) = \sum_{n=0}^{\infty} \text{Pr}^{\text{extended}}\{n\} s^n \qquad (5.87)$$

Denote the conditional probability function of the spatial distribution of a sensor measurement by $p(Y_i \,|\, X_\ell)$. Then the PGF of the histogram of the number of points generated by a single target in state cell X_ℓ by

$$G_\ell^{\text{extended}}(y_1, \dots, y_m) = G_{\text{extended}}^{\#}\left(\sum_{i=1}^{m} p(Y_i \,|\, X_\ell) y_i\right) \qquad (5.88)$$

The PGF of the target-measurement histogram cell counts is identical to (5.72) except that it uses the extended target model (5.88) instead of point target model $G_\ell(\cdot)$ in (5.61).

The difficulty encountered by this approach is that the derivatives of the joint PGF needed to compute the posterior distribution via Bayes' theorem quickly become unwieldy for manual evaluation. This is an active area of research. See, for example, [11] and the references therein.

Other problems share the same difficulty. The multisensor problem is one. For further comments on this topic, see Section 5.7.1.

5.6 SUMMARY

Counting the number of targets that are present in a surveillance region is a difficult problem. Estimating the number of targets per unit state space is the goal of the filters discussed in this chapter, not estimating the multitarget state.

We began by using finite point processes to model multitarget state. After reviewing PPPs and their basic properties, we used an augmented state space S^+ to derive the iFilter and PHD filter. This had several advantages. First, and most importantly, we used the augmented state space to finesse the all-important measurement-to-target association problem. Measurement-to-target assignments are not avoided—they are accounted for explicitly in the derivation of the iFilter, but explicit computation of associations (or association probabilities) of measurements to targets do not appear in the iFilter recursion. Second, we were able to derive the iFilter and the PHD filter from the augmented state space formulation. We showed that the output of these filters is the intensity function of a PPP, and this fact provided insight into the nature of the multitarget state model.

Using the augmented state space enabled the combinatorial character of the assignment problem to be treated explicitly. Alternatively, without adopting an augmented state space, we can avoid working directly with the assignments by working instead with the PGFL of the finite point process model of multitarget state [2, 3]. This approach embeds, or encodes, the combinatorial models of the multitarget estimation problem into the PGFL. The PGFL thus completely characterizes the estimation problem. This approach has the virtue of reducing intensity filter derivation to the calculation of functional derivatives of the PGFL. In effect, functional differentiation decodes the PGFL. The PGFL approach can be extended to augmented spaces; for details, see [12].

Instead of using the PGFL approach, we have followed a different path, a mathematical path laid out by Moyal in his fundamental 1962 paper on point processes [5]. This starts with classical PGFs (i.e., z-transforms) and discrete random variables. Our random variables are the target and measurement counts in discretized grids in target and measurement space. Bayesian analysis leads to estimates of the numbers of targets in the target space grid cells, conditioned on the observed numbers of measurements in the measurement space grid cells. Taking small cell limits yields the iFilter and PHD filter. Proceeding in this natural way establishes that the multitarget models of these filters are spatial counting models. This is an important fact, one not evident from the PGFL approach.

The PGF approach has the virtue of exposing the nature of the multitarget modeling. In effect, it shows that extracting target state estimates from the summary statistics (e.g., the intensity function) of the Bayes posterior point process is intuitively reasonable but not an optimal procedure. For optimal state estimation, it is necessary to use a method that estimates target states directly. The PGF approach firmly establishes that the multitarget model is an example of what are called spatial branching processes in the mathematical and physics literatures. Branching processes [5, 13, 14] have long been used in biology to model population processes over space and time, and in physics to model chain reactions and neutron transport. This is satisfying because filters such as the iFilter and PHD can be better understood by seeing how they are related to problems in other applications.

5.7 NOTES

We end with a discussion of some topics not covered in the chapter and provide some background on the development of the methods used in the chapter.

5.7.1 Other Topics

Several topics are not discussed. Non-Poisson distributions of the numbers of targets and clutter is one topic. The form of the joint PGF given in (5.72) is very general—any desired distribution of the numbers of targets and clutter can be considered. These distributions have to be information-updated and approximated to maintain a closed recursion. The small cell limit can then be investigated if desired. The cardinalized PHD (CPHD) filter [3] posits a finite maximum number of targets and computes approximations to the information updates of the distribution of the total number (i.e., the cardinal, or canonical, number) of targets and their spatial distribution. It does not however use discrete cells, and so works with PGFLs instead of PGFs.

Multisensor problems are also not discussed. The PGF for these problems can be formulated but, like the extended target problem, the derivatives needed to find the intensity function of the Bayes posterior target distribution are not conducive to manual evaluation. One way around the problem is to redefine it. Thus, instead of counting the number of targets per unit state space, we can count the number of sensor target detections per unit state space. This changes the PGF of the multisensor problem to a form whose derivatives can be evaluated manually. The resulting filter is called the traffic filter, and it is very closely related to the medical imaging technology called single-photon emission computed tomography (SPECT). For further details on the traffic filter, see [15]. For discussion of the close connections between SPECT and traffic filters, and between intensity filters and positron emission tomography (PET), see [1, 16].

Section 5.2.6 mentions the correlation between points of the Bayes posterior point process. Correlation between different points in a continuous space, and between the numbers of targets in different cells in a discrete state space, is an interesting and perhaps surprising phenomenon given the a priori assumption that targets are mutually independent. It is sometimes referred to as "spooky action at a distance" [17] because of its similarity to problems in physics. The target-measurement PGFL (or the discrete-cell PGF) can be used to derive explicit formulae for the pair-correlation function and the reduced Palm intensity function of the Bayes posterior process [7]. It is also noted in [7] that explicit three-way and higher order correlation functions can be derived. The utility of these and related functions for tracking is largely unexplored.

An important topic is what might be called a "batch intensity filter" for a sequence of measurement scans. One way to address such multiscan problems is to find the appropriate PGF and take the small cell limit. The PGF can be found via a backward recursion of depth equal to the batch length. This is ongoing work, but the necessary recursion can be found by following the methods presented in [18, 19].

5.7.2 Background

Branching processes and finite point processes are established topics in probability and statistics. PGFLs for finite point processes were introduced in 1962 by Moyal [5]. In this paper Moyal noted the connection between the PGFs of discrete random variables and PGFLs, as well as the connections to branching processes. For detailed discussions of branching processes, see the well-known texts [13, 14].

Branching processes and point process theory were studied extensively by Harris in 1963 [14]. Daley and Vere-Jones have written the modern authoritative texts [20, 21] on point process theory. They write [21, p. 1] that point process theory "reached a definitive form in the now classic treatments by Moyal (1962) and Harris (1963)."

Mahler applied PGFLs to multitarget tracking problems in a series of papers; see [2, 3] and the references therein. In this corpus he used a finite set statistics (FISST) calculus to derive the PHD and CPHD filters. He introduced random finite set (RFS) models for the multitarget state, and the idea of recursively approximating the Bayes posterior point process by what we call in this chapter a PPP. The term PHD was coined by Stein and Winter [22], who viewed the process of evidence accrual as additive, as opposed to multiplicative. The reformulation of the PHD using RFSs is due to Mahler [23].

The first papers to provide an alternative derivation of the PHD filter are the series of papers by Erdinc, Willett, and Bar-Shalom [24–26]. Their physical-space approach is intuitively appealing in that it models flow between bins in a discretized model of target state space and then recovers the PHD filter in the limit as bin size goes to zero. Their approach differs considerably from the PGF approach used here. These papers also derive the CPHD filter, and they compare and contrast these filters with other approaches to multitarget tracking.

References

[1]   Streit, R., *Poisson Point Processes—Imaging, Tracking, and Sensing*, New York: Springer, 2010.

[2]   Mahler, R. P. S., "Multitarget Bayes filtering via first-order multitarget moments," *IEEE Transactions Aerospace and Electronic Systems*, Vol. AES-39, 1152-1178, 2003.

[3]   Mahler, R. P. S., *Statistical Multisource-Multitarget Information Fusion*, Norwood, Massachusetts: Artech House, 2007.

[4]   Kingman, J., *Poisson Processes*, Oxford: Oxford University Press, 1993.

[5]   Moyal, J. E., "The general theory of stochastic population processes," *Acta Mathematica*, Vol. 108, 1-31, 1962.

[6]   Jaacola, T., *Machine Learning: Discriminative and Generative*, Boston: Kluwer Academic Publishers, 2004.

[7]   Bozdogan, Ö., Efe, M., and Streit, R., "Reduced Palm intensity for track extraction," *Proceedings of the 16th ISIF International Conference on Information Fusion*, Istanbul, July 2013.

[8]   Cressie, N., and Wikle, C. K., *Statistics for Spatio-Temporal Data*, Hoboken, New Jersey: Wiley & Sons, 2011.

[9]   Wilf, H. S., *generatingfunctionology*, Academic Press, Second Edition, 1994. (Open access internet edition, http://www.math.upenn.edu/~wilf/DownldGF.html)

[10] Chen, X., Tharamrasa, R., Kirubarajan, T., and Pelletier, M., "Integrated clutter estimation and target tracking using Poisson point process," *Proceedings of the SPIE Conference on Signal and Data Processing of Small Targets*, San Diego, CA, SPIE Vol. 7445, 2009.

[11] Orguner, U., Lundquist, C., and Granström, K., "*Extended target tracking with a cardinalized probability hypothesis density filter,*" Dept. of Electrical Engineering, Linköpings Universitet, Technical Report LiTH-ISY-R-2999, 14 March 2011.

[12] Streit, R., "The probability generating functional for finite point processes, and its application to the comparison of PHD and intensity filters," *Journal of Advances in Information Fusion*, Vol. 8, No. 2, December 2013.

[13] Athreya, K. B., and Ney, P. E., *Branching Processes*, Berlin: Springer-Verlag, 1972. (Republished by Mineola, New York: Dover, 2004.)

[14] Harris, T. E., *The Theory of Branching Processes*, Berlin: Springer-Verlag, 1963.

[15] Streit, R., "Multisensor traffic mapping filters," *Workshop on Sensor Data Fusion*, Bonn, Germany, 4-6 September 2012.

[16] Streit, R. L., "PHD intensity filtering is one step of an EM algorithm for positron emission tomography," *Proceedings of the 12th International Conference on Information Fusion*, Seattle, July 2009.

[17] Fränken, D., Schmidt, M., and Ulmke, M., "'Spooky action at a distance' in the cardinalized probability hypothesis density filter," *IEEE Transactions on Aerospace and Electronic Systems*, Vol. AES-45, 1657-1664, 2009.

[18] Streit, R., "How to count targets given only the number of measurements," *Proceedings of the 16th ISIF International Conference on Information Fusion*, Istanbul, July 2013.

[19] Streit, R., "Intensity filters on discrete spaces," *IEEE Transactions on Aerospace and Electronic Systems*, to be published.

[20] Daley, D. J., and Vere-Jones, D., *An Introduction to the Theory of Point Processes, Vol. I: Elementary Theory and Methods*, New York: Springer, 1988 (Second Edition, 2003).

[21] Daley, D. J., and Vere-Jones, D., *An Introduction to the Theory of Point Processes, Vol. II: General Theory and Structure*, New York: Springer, 1988 (Second Edition, 2008).

[22] Stein, M. C., and Winter, C. L., "*An additive theory of probabilistic evidence accrual,*" Report LA-UR-93-3336, Los Alamos National Laboratory, 1993.

[23] Mahler, R. P. S., "A theoretical foundation for the Stein-Winter probability hypothesis density (PHD) multitarget tracking approach," *Proceedings of the MSS National Symposium on Sensor and Data Fusion*, Vol. 1, 99-117, San Antonio, TX, 2000.

[24] Erdinc, O., Willett, P., and Bar-Shalom, Y., "Probability hypothesis density filter for multitarget multisensor tracking," *Proceedings of the 8th ISIF International Conference on Information Fusion*, Philadelphia, July 2005.

[25] Erdinc, O., Willett, P., and Bar-Shalom, Y., "A physical-space approach for the probability hypothesis density and cardinalized probability hypothesis density filters," *Proceedings of the SPIE Conference on Signal and Data Processing of Small Targets*, SPIE Vol. 6236, Orlando, FL, April 2006.

[26] Erdinc, O., Willett, P., and Bar-Shalom, Y., "A physical-space approach for the probability hypothesis density and cardinalized probability hypothesis density filters," *IEEE Transactions on Aerospace and Electronic Systems*, Vol. AES-45, 1657–1664, October 2009.

Chapter 6

Multiple Target Tracking Using Tracker-Generated Measurements

Chapter 4 develops classical Bayesian multiple target tracking by dividing the problem into three steps: generation of contacts from sensor data, association of contacts to targets, and target state estimation. We assume that the contact generation step has already been performed and consider only the contact association and target state estimation steps. If the association of contacts to targets were always unambiguous, we could decompose the multiple target tracking problem into independent single target problems and proceed as in Chapter 3. When association is ambiguous, the tracking and association problems become coupled and difficult to solve. In Chapters 4 and 5, we develop several methods for addressing this problem.

In Chapter 4, we define a contact as an observation that consists of a called detection and a measurement of one or more target parameters. Contacts are usually obtained from a sensor by applying a threshold to the sensor responses to declare detections and then performing signal processing on the responses that cross the threshold to obtain the measurements. This chapter drops the assumption that the observations provided to the tracker are in the form of contacts and assumes that we have access to unthresholded sensor data. We develop the maximum a posteriori probability penalty function (MAP-PF) technique [1–5] for Bayesian multiple target tracking, which jointly performs target state estimation and the signal processing required to obtain target measurements.

The MAP-PF approach is formulated as a target state point estimation problem using the MAP estimation criterion; however, the solution requires computation of the posterior distribution of the target state. Thus the MAP-PF technique is fundamentally a Bayesian filtering technique that subsequently produces the MAP point estimate. The solution to the MAP-PF problem is a two-stage estimation process that consists of the extraction of one measurement for each target from the sensor data followed by the estimation of target state using

Bayesian filtering. This two-stage process is similar to the three-stage classical methods. The critical difference is that in MAP-PF, the measurement generation and target state estimation stages are coupled by the penalty function and performed by the tracker, while the association step of classical methods is eliminated.

6.1 MAXIMUM A POSTERIORI PENALTY FUNCTION TRACKING

We now develop the MAP-PF approach to multiple target tracking. The description of the problem and approach taken are motivated by the problem of tracking acoustic targets with linear arrays of passive acoustic sensors (hydrophones). The reader will find it useful to keep this example in mind as he reads this chapter. However, the MAP-PF approach is not limited to this case as will be discussed in Section 6.6.

Array responses are typically produced by performing Fourier analysis on a segment of time series data from array elements to compute complex Fourier coefficients of the frequency response at the elements. This corresponds to a scan of data. These responses can be processed (beamformed) to produce angle-of-arrival estimates of the energy sources within range of the array. An overview of the sensor array observation model and the signal processing required to obtain sensor measurements is provided in Section 6.7. This places us in a situation similar to the bearings-only tracking example (example 2) in Chapter 1 where the target state space consists of the usual spatial coordinates, and the sensor measurements (bearings) are nonlinear functions of target state. There is a crucial difference from the example in Chapter 1. In that example, we assume that the sensor responses were thresholded and analyzed to produce line of bearing contacts. In this chapter we will be dealing with unthresholded sensor data, and the data used to compute the sensor response will have signals from all targets superimposed on one other.

For MAP-PF, we follow the multiple target tracking assumptions of Section 4.1. However, we assume that the number N of targets is known and does not change over the scan times $t_1, \ldots, t_K$. In practice this means that we must employ an auxiliary process such as the likelihood ratio detector tracker (LRDT) described in Chapter 7 to determine the number of targets present. In contrast to the scan assumptions in Section 4.2, we assume that a scan is a sensor response at a given time that contains contributions from all targets simultaneously.

As in Section 4.1, we assume that the multiple target likelihood functions at distinct times are independent given system state s. That is, if $\mathbf{y}_K = (y_1, \ldots, y_K)$ are the sensor responses at times $t_1, \ldots, t_K$ and $\mathbf{s}_1, \ldots, \mathbf{s}_K$ the system states, then

$$L(\mathbf{y}_K \mid \mathbf{s}_1,...,\mathbf{s}_K) = \prod_{k=1}^{K} L_k(y_k \mid \mathbf{s}_k) \text{ for } \mathbf{s}_k \in \mathbf{S} \equiv S \times \cdots \times S \qquad (6.1)$$

where the product defining $\mathbf{S}$ is taken N times. We do not assume that the scan likelihood function $L_k(y_k \mid \mathbf{s}_k)$ can be factored into conditionally independent likelihood functions for each target.

Following Section 4.3, we assume that the Markov motion processes for the individual targets are mutually independent so that the multiple target transition function factors as follows:

$$q_k(\mathbf{s}_k \mid \mathbf{s}_{k-1}) = \prod_{n=1}^{N} q_k(s_{n,k} \mid s_{n,k-1}, n) \qquad (6.2)$$

where $\mathbf{s}_k = (s_{1,k},...,s_{N,k})$ and $\mathbf{s}_{k-1} = (s_{1,k-1},...,s_{N,k-1})$. We also assume that the posterior distributions on the state of target n at time t_{k-1} for $n = 1,...,N$, are mutually independent so that the posterior distribution has the form

$$p(t_{k-1}, \mathbf{s}_{k-1}) = \prod_{n=1}^{N} p_n(t_{k-1}, s_{n,k-1}) \qquad (6.3)$$

As a result the posterior distribution on system state at time t_k can be written as

$$
\begin{aligned}
p(t_k, \mathbf{s}_k) &= \frac{1}{C} L_k(y_k \mid \mathbf{s}_k) \int q_k(\mathbf{s}_k \mid \mathbf{s}_{k-1}) p(t_{k-1}, \mathbf{s}_{k-1}) \, d\mathbf{s}_{k-1} \\
&= \frac{1}{C} L_k(y_k \mid \mathbf{s}_k) \prod_{n=1}^{N} \left[\int q_k(s_{n,k} \mid s_{n,k-1}, n) p_n(t_{k-1}, s_{n,k-1}) \, ds_{n,k-1} \right]
\end{aligned}
\qquad (6.4)
$$

where C is a normalization constant.

6.1.1 MAP-PF Formulation

MAP-PF seeks to find the MAP set of paths for the N targets given a set of K scans of data (as defined above) from the sensor. In this section we describe the mathematical formulation of this problem.

6.1.1.1 MAP Target State Estimation

The MAP target state estimate $\hat{\mathbf{s}}_k$ at time t_k is found from:

$$\hat{\mathbf{s}}_k = \arg\max_{\mathbf{s}_k} \left\{ \ln p(t_k, \mathbf{s}_k) \right\} \tag{6.5}$$

Applying the logarithm to the first equation in (6.4), we have

$$\ln p(t_k, \mathbf{s}_k) = -\ln C + \ln L_k(y_k \,|\, \mathbf{s}_k) + \ln \int q_k(\mathbf{s}_k \,|\, \mathbf{s}_{k-1}) p(t_{k-1}, \mathbf{s}_{k-1}) d\mathbf{s}_{k-1}$$

In performing the maximization, we can ignore the constant term, so the problem is to find the MAP target state estimate by solving the following optimization problem:

$$\hat{\mathbf{s}}_k = \arg\max_{\mathbf{s}_k} \left\{ \ln L_k(y_k \,|\, \mathbf{s}_k) + \ln \int q_k(\mathbf{s}_k \,|\, \mathbf{s}_{k-1}) p(t_{k-1}, \mathbf{s}_{k-1}) d\mathbf{s}_{k-1} \right\} \tag{6.6}$$

This is a difficult multidimensional, nonlinear optimization problem. In order to facilitate the solution to (6.6) we introduce an "auxiliary variable" θ to represent the "natural" measurement space of the sensor. For example line arrays usually produce a bearing measurement θ, which is a nonlinear function $H(\cdot)$ of the target state. The introduction of the auxiliary parameter θ transforms the unconstrained optimization problem in (6.6) into an equivalent constrained optimization problem in which we estimate both the target state and the auxiliary parameter. Initially this makes the problem seem more complicated but ultimately it allows us to decompose the problem into the solution of a sequence of simpler unconstrained optimization problems using the penalty function method. We will see that estimating θ for each target corresponds to generating a measurement for each target from the sensor data, with the penalty function representing a joint measurement likelihood function that can be factored into independent measurement likelihood functions.

6.1.1.2 Constrained Optimization Problem with Auxiliary Variables

Suppose that $\theta(\mathbf{s})$ is the set of parameters measured by the sensor due to system state $\mathbf{s} \in \mathbf{S}$, e.g., the vector of bearings to the array of the targets located at $\mathbf{s} = (s_1, \ldots, s_N)$. The sensor measurement parameters are related to the system state by the nonlinear transformation

$$\theta(\mathbf{s}) = \mathbf{H}(\mathbf{s}) = \left(H(s_1), \ldots, H(s_N) \right) \text{ for } \mathbf{s} = \left(s_1, \ldots, s_N \right) \in \mathbf{S}$$

Let $s_{n,k}$ be the state of the nth target at time t_k. Define

$$\theta_{n,k} = H\left(s_{n,k}\right) \tag{6.7}$$

and let $\boldsymbol{\theta}_k = \left(\theta_{1,k}, \dots, \theta_{N,k}\right)$.

We assume that for a scan of data y_k at time t_k, the likelihood function depends on the target state only through the transformation $H(s_k)$, that is, $L_k(y_k \mid s_k) = L_k(y_k \mid H(s_k))$. Then since $\boldsymbol{\theta}_k = H(s_k)$, the likelihood function can be expressed in terms of the auxiliary variables as follows:

$$\begin{aligned} L_k(y_k \mid s_k) &= L_k\left(y_k \mid H(s_k)\right) \\ &= L_k\left(y_k \mid \boldsymbol{\theta}_k\right) \end{aligned} \tag{6.8}$$

The likelihood expressions in (6.8) provide a mathematical representation of the notion of the "natural" measurement space of the sensor, (i.e., that the likelihood function depends on the target state only through a transformation to these parameters).

We now restate the optimization problem in (6.6) in terms of both the original state variables s_k and the new auxiliary variables $\boldsymbol{\theta}_k$ as follows:

$$\hat{s}_k, \hat{\boldsymbol{\theta}}_k = \arg\max_{s_k, \boldsymbol{\theta}_k} \left\{ \ln L_k(y_k \mid \boldsymbol{\theta}_k) + \ln \int q_k(s_k \mid s_{k-1}) p(t_{k-1}, s_{k-1}) \, ds_{k-1} \right\} \tag{6.9}$$

$$\text{subject to } \boldsymbol{\theta}_k = H(s_k)$$

By enforcing the relationship $\boldsymbol{\theta}_k = H(s_k)$ the new constrained optimization problem in (6.9) is equivalent to the original unconstrained optimization problem in (6.6).

6.1.1.3 MAP-PF Optimization Problem

The formulation in (6.9) allows us to use the classical penalty function method for nonlinear constrained optimization problems [6, 7]. The penalty function method is an iterative procedure that involves solving a sequence of (easier) unconstrained optimization problems that converge to the original constrained optimization problem. The unconstrained problems are related to the original constrained problem by the addition of a continuous, differentiable penalty function that is equal to zero in the region where the constraint is satisfied (the feasible region) and that is negative outside that region (the infeasible region). In this problem the feasible region is:

$$\Omega = \left\{ (s_k, \boldsymbol{\theta}_k) : \boldsymbol{\theta}_k = H(s_k) \right\}$$

We define a penalty function $\Phi_k\left(\boldsymbol{\theta}_k, \mathbf{H}(\mathbf{s}_k)\right)$ that has the properties described above; that is,

$$\Phi_k\left(\boldsymbol{\theta}_k, \mathbf{H}(\mathbf{s}_k)\right) = 0 \quad \boldsymbol{\theta}_k = \mathbf{H}(\mathbf{s}_k)$$
$$\Phi_k\left(\boldsymbol{\theta}_k, \mathbf{H}(\mathbf{s}_k)\right) < 0 \quad \boldsymbol{\theta}_k \neq \mathbf{H}(\mathbf{s}_k)$$

The strength of the penalty function is controlled by a sequence of positive, decreasing penalty parameters $c_m; m = 1, \ldots, \infty$. We now replace (6.9) with a sequence of optimization problems where the mth problem in the sequence is to find

$$\hat{\mathbf{s}}_k^m, \hat{\boldsymbol{\theta}}_k^m = \underset{\mathbf{s}_k, \boldsymbol{\theta}_k}{\arg \max} \left\{ \begin{array}{l} \ln L_k(y_k \mid \boldsymbol{\theta}_k) + \ln \int q_k(\mathbf{s}_k \mid \mathbf{s}_{k-1}) p(t_{k-1}, \mathbf{s}_{k-1}) d\mathbf{s}_{k-1} \\ \qquad\qquad\qquad\qquad\qquad + c_m^{-1} \Phi_k\left(\boldsymbol{\theta}_k, \mathbf{H}(\mathbf{s}_k)\right) \end{array} \right\} \qquad (6.10)$$

The penalty function method is beneficial if the optimization problem in (6.10) is easier to solve than the original problem in (6.9) and the solution at the mth iteration can be used to initialize the solution at the $(m+1)$th iteration. Initially, the penalty is quite loose and infeasible solutions are obtained. With each iteration, the penalty term is forced closer to the "ideal" penalty function, which is equal to negative infinity in the infeasible region; the unconstrained problem is forced closer to the original constrained problem; and the solution is forced into the feasible region. The optimization problems, as well as their solutions, converge in the limit. An overview of the penalty method and a proof of convergence to a stationary point is provided in [7].

6.1.1.4 Quadratic Penalty Function

For our problem it is particularly convenient to choose the following quadratic penalty function,

$$\Phi_k\left(\boldsymbol{\theta}_k, \mathbf{H}(\mathbf{s}_k)\right) = \sum_{n=1}^{N} \phi_{n,k}\left(\theta_{n,k}, H(s_{n,k})\right)$$
$$\phi_{n,k}\left(\theta, H(x)\right) = -\frac{1}{2}\left(\theta - H(x)\right)^T \Sigma_{n,k}^{-1}\left(\theta - H(x)\right) \qquad (6.11)$$

where $\Sigma_{n,k}$ is a parameter that defines the shape of the quadratic function $\phi_{n,k}\left(\theta, H(x)\right)$. In the case where θ is a scalar measurement such as bearing, $\Sigma_{n,k}$ is a positive number. More generally, if θ is a vector then $\Sigma_{n,k}$ is a positive definite matrix.

The quadratic penalty function is chosen because of its relationship to the Gaussian probability density function. Let Σ_k denote the block diagonal matrix with main diagonal blocks $(\Sigma_{1,k}, \ldots, \Sigma_{N,k})$. The exponential of the quadratic function $c_m^{-1}\phi_{n,k}(\theta, H(x))$ is proportional to a Gaussian probability density function with mean $H(x)$ and covariance matrix $c_m\Sigma_{n,k}$. Thus the exponential of the penalty function weighted by the penalty parameter is proportional to a Gaussian probability density function with mean $\mathbf{H}(\mathbf{s}_k)$ and covariance matrix $c_m\Sigma_k$,

$$
\begin{aligned}
\exp&\left\{c_m^{-1}\Phi_k\left(\boldsymbol{\theta}_k, \mathbf{H}(\mathbf{s}_k)\right)\right\} \\
&= \exp\left\{c_m^{-1}\sum_{n=1}^{N} -\frac{1}{2}\left(\theta_{n,k} - H(s_{n,k})\right)^T \Sigma_{n,k}^{-1}\left(\theta_{n,k} - H(s_{n,k})\right)\right\} \\
&\propto \prod_{n=1}^{N} \eta\left(\theta_{n,k}, H(s_{n,k}), c_m\Sigma_{n,k}\right) \\
&\propto \eta\left(\boldsymbol{\theta}_k, \mathbf{H}(\mathbf{s}_k), c_m\Sigma_k\right)
\end{aligned}
\tag{6.12}
$$

Equivalently, the weighted penalty function is equal to the logarithm of the Gaussian probability density plus a constant

$$
\begin{aligned}
c_m^{-1}\Phi_k\left(\boldsymbol{\theta}_k, \mathbf{H}(\mathbf{s}_k)\right) &= \sum_{n=1}^{N} \ln \eta\left(\theta_{n,k}, H(s_{n,k}), c_m\Sigma_{n,k}\right) + C \\
&= \ln \eta\left(\boldsymbol{\theta}_k, \mathbf{H}(\mathbf{s}_k), c_m\Sigma_k\right) + C
\end{aligned}
\tag{6.13}
$$

6.1.2 Iterative Optimization

The introduction of the auxiliary parameter $\boldsymbol{\theta}_k$ and the formulation of the optimization problem as the solution of a sequence of unconstrained problems with a carefully chosen penalty function allows us to decompose the problem into the solution of a sequence of simpler optimization problems. In particular the unconstrained problem in (6.10) can be solved iteratively using cyclic maximization by fixing the target state parameters and maximizing over the auxiliary parameters and then holding the auxiliary parameters fixed and maximizing over the target state parameters. We shall see below that finding the auxiliary parameters $\boldsymbol{\theta}_k^m$ that maximize (6.10) while holding $\mathbf{s}_k$ fixed is equivalent to generating a measurement for each of the targets at time t_k and that the exponential of the penalty function in (6.11) can be interpreted as a product of Gaussian likelihood functions where the quantity $c_m\Sigma_{n,k}$ represents the variance of the error in obtaining the measurement $\hat{\theta}_{n,k}^m$ from the sensor data y_k.

6.1.2.1 Maximize over Auxiliary Parameters

Since the second term in the brackets in (6.10) does not depend on $\boldsymbol{\theta}_k$, the MAP auxiliary parameter estimate at time t_k given $\mathbf{s}_k$ is found by maximizing:

$$\hat{\boldsymbol{\theta}}_k^m = \arg\max_{\boldsymbol{\theta}_k} \left\{ \ln L_k(y_k \mid \boldsymbol{\theta}_k) + c_m^{-1} \Phi_k(\boldsymbol{\theta}_k, \mathbf{H}(\mathbf{s}_k)) \right\} \qquad (6.14)$$

This is a penalized maximum likelihood estimation problem. If we define $c_m^{-1}\Phi_k(\boldsymbol{\theta}_k, \mathbf{H}(\mathbf{s}_k)) = \ln p(\boldsymbol{\theta}_k) + C$, it can also be interpreted as a MAP estimation problem with prior density $p(\boldsymbol{\theta}_k) = \eta(\boldsymbol{\theta}_k, \mathbf{H}(\mathbf{s}_k), c_m \Sigma_k)$. It is a multidimensional nonlinear optimization problem that requires joint maximization over the N components of $\boldsymbol{\theta}_k$.

This problem can also be solved iteratively using cyclic maximization by maximizing over the auxiliary parameters one at a time, holding the remaining N-1 auxiliary parameters fixed. Then for each (n,k) we solve for $\hat{\theta}_{n,k}^m$ as follows:

$$\hat{\theta}_{n,k}^m = \arg\max_{\theta} \left[\begin{array}{l} \ln L_k(y_k \mid \theta_{1,k}, \ldots, \theta_{n-1,k}, \theta, \theta_{n+1,k}, \ldots, \theta_{N,k}) \\ \qquad\qquad + c_m^{-1}\phi_{n,k}(\theta, H(s_{n,k})) \end{array} \right] \qquad (6.15)$$

6.1.2.2 Maximize over Target State

Similarly, since the first term in the brackets in (6.10) does not depend on $\mathbf{s}_k$, the MAP state vector estimate at time k given $\boldsymbol{\theta}_k$ is found by maximizing:

$$\begin{aligned} \hat{\mathbf{s}}_k^m &= \arg\max_{\mathbf{s}_k} \left\{ \ln \int q_k(\mathbf{s}_k \mid \mathbf{s}_{k-1}) p(t_{k-1}, \mathbf{s}_{k-1}) d\mathbf{s}_{k-1} + c_m^{-1}\Phi_k(\boldsymbol{\theta}_k, \mathbf{H}(\mathbf{s}_k)) \right\} \\ &= \arg\max_{\mathbf{s}_k} \left\{ \ln p^-(t_k, \mathbf{s}_k) + \ln \eta(\boldsymbol{\theta}_k, \mathbf{H}(\mathbf{s}_k), c_m \Sigma_k) + C \right\} \\ &= \arg\max_{\mathbf{s}_k} \left\{ \ln p(t_k, \mathbf{s}_k) \right\} \end{aligned} \qquad (6.16)$$

where

$$\begin{aligned} p(t_k, \mathbf{s}_k) &= \frac{1}{C'} p^-(t_k, \mathbf{s}_k) \eta(\boldsymbol{\theta}_k, \mathbf{H}(\mathbf{s}_k), c_m \Sigma_k) \\ &= \frac{1}{C''} p^-(t_k, \mathbf{s}_k) \exp\left\{ c_m^{-1}\Phi_k(\boldsymbol{\theta}_k, \mathbf{H}(\mathbf{s}_k)) \right\} \end{aligned} \qquad (6.17)$$

Thus to obtain the MAP estimate, we must first compute the posterior density in (6.17). This can be accomplished as in Section 4.1 with recursive Bayesian filtering consisting of a standard motion update followed by an information update with the exponential of the weighted penalty function taking the place of the scan likelihood function.

We purposely specified the penalty function in (6.11) to be a sum of quadratic functions that each depend only on the state vector and auxiliary variable for a single target. Then the exponential of the weighted penalty function is a product of Gaussian likelihood functions as shown in (6.12). Combined with the independent motion model assumption in (6.2), the posterior distribution of the joint target state vector meets the requirements of the independence theorem in Section 4.3 and can be expressed as a product of independent posterior target state distributions

$$p(t_k, \mathbf{s}_k) = \prod_{n=1}^{N} p_n(t_k, s_{n,k}) \tag{6.18}$$

where

$$p_n(t_k, s_{n,k}) = \frac{1}{C} \exp\left\{ c_m^{-1} \phi_{n,k}\left(\theta_{n,k}, H(s_{n,k}) \right) \right\} p_n^-(t_k, s_{n,k}) \tag{6.19}$$

and

$$p_n^-(t_k, s_{n,k}) = \int q_k(s_{n,k} \mid s_{n,k-1}, n) p_n(t_{k-1}, s_{n,k-1}) ds_{n,k-1} \tag{6.20}$$

Then the MAP optimization problem in (6.16) becomes

$$\hat{\mathbf{s}}_k^m = \arg\max_{\mathbf{s}_k} \left\{ \sum_{n=1}^{N} \ln p_n(t_k, s_{n,k}) \right\} \text{ where } \mathbf{s}_k = \left(s_{1,k}, \ldots, s_{N,k} \right) \tag{6.21}$$

which can be solved by performing N separate maximizations of the form

$$\hat{s}_{n,k}^m = \arg\max_x \left\{ \ln p_n(t_k, x) \right\} \tag{6.22}$$

If $p_n^-(t_k, \cdot)$ is a Gaussian density, then $p_n(t_k, \cdot)$ is also Gaussian, and $\hat{s}_{n,k}^m$ is equal to the mean of that density.

Thus the Bayesian filtering recursions can be performed separately for each target, and the posterior distribution in (6.18) meets the independence assumption in (6.3) for the next scan. Given the interpretation of the exponential of the penalty function as the product of independent Gaussian likelihood functions, $\Sigma_{n,k}$ represents the variance of the error in obtaining an estimate of $\theta_{n,k}$ from the sensor data y_k. One may choose $\Sigma_{n,k}$ using the same rationale used for specifying the measurement variance in a standard Bayesian filter. One method for doing this is to compute the Cramér-Rao bound (CRB), which is a lower bound on the variance of any unbiased estimator and is achievable asymptotically by the maximum likelihood estimator [8]. Typically, one would compute the CRB for jointly estimating all components of θ_k and set $\Sigma_{n,k}$ equal to the nth diagonal component. Ignoring any nonzero off-diagonal components in the CRB is an approximation that allows us to treat the measurements as being independent.

6.1.2.3 Iterations, Convergence, and Practical Implementation

The solution to the MAP-PF optimization problem described above involves three levels of iteration. The outer level is the iteration over m in the penalty function optimization problem in (6.10) in which the penalty weighting c_m^{-1} is made stronger at each iteration. For each of these iterations, there is an iteration between auxiliary parameter estimation in (6.14) and target state estimation in (6.16). The auxiliary parameter estimation step involves another iteration over its N components in (6.15). In theory, one repeats each iteration until a convergence criterion is met. In practice, multiple iterations at each level would make the implementation computationally infeasible in most applications. Fortunately, satisfactory results may be obtained using one iteration at each step. In the algorithm summary provided in Section 6.1.3, we provide a practical implementation using one iteration at the outermost level ($m = 1$ with $c_1 = 1$) and one iteration at each of the inner levels.

To perform the optimizations in (6.15) and the information updates in (6.19) we must substitute current point estimates for the parameters that we have assumed to be fixed and known. In (6.19), we need an estimate for the auxiliary parameter $\theta_{n,k}$. This is obtained from the auxiliary parameter estimation step in (6.15). In (6.15), we need estimates for the target state $s_{n,k}$ as well as the remaining N-1 auxiliary parameters. At the time that the sensor data is collected, the best estimate of the target state can be obtained from the motion-updated target state distribution $p_n^-(t_k, s_{n,k})$. For consistency with the MAP formulation, we specify the estimate as the maximum of this distribution, although any suitable estimate may be used. We denote this estimate as $\hat{s}_{n,k}^-$. Auxiliary parameter estimates are obtained from the transformation $H(\hat{s}_{n,k}^-)$.

6.1.3 Algorithm

Let $q_0(x \mid n) = \Pr\{X_n(0) = x\}$ and $q_k(x \mid x', n) = \mathbf{Pr}\{X_n(t_k) = x \mid X_n(t_{k-1}) = x'\}$. Under the assumptions given above, we now state the MAP-PF recursion that solves the MAP-PF estimation problem with one iteration step at each of the iteration levels.

MAP-PF Recursion

Let y_k be the sensor response in the kth scan. Set $t_0 = 0$ and

$$p_n(t_0, x) = q_0(x \mid n) \text{ for } x \in S \text{ and } n = 1, \ldots, N$$

For $k = 1, 2, \ldots$ do the following:

Motion update. For $n = 1, \ldots, N$

$$p_n^-(t_k, x) = \int_S q_k(x \mid x', n) \, p_n(t_{k-1}, x') \, dx' \tag{6.23}$$

Predicted target state estimate. For $n = 1, \ldots, N$

$$\hat{s}_{n,k}^- = \arg\max_{x \in S} p_n^-(t_k, x) \tag{6.24}$$

Measurements. For $n = 1, \ldots, N$

$$\phi_{n,k}\left(\theta, H(\hat{s}_{n,k}^-)\right) = -\frac{1}{2}\left(\theta - H(\hat{s}_{n,k}^-)\right)^T \Sigma_{n,k}^{-1}\left(\theta - H(\hat{s}_{n,k}^-)\right)$$

$$\hat{\theta}_{n,k} = \arg\max_{\theta}\left\{ \begin{array}{l} \ln L_k\left(y_k \mid H(\hat{s}_{n,1}^-), \ldots, H(\hat{s}_{n-1,k}^-), \theta, H(\hat{s}_{n+1,k}^-), \ldots, H(\hat{s}_{N,k}^-)\right) \\ + \phi_{n,k}\left(\theta, H(\hat{s}_{n,k}^-)\right) \end{array} \right\} \tag{6.25}$$

Information update. For $n = 1, \ldots, N$

$$\phi_{n,k}\left(\hat{\theta}_{n,k}, H(x)\right) = -\frac{1}{2}\left(\hat{\theta}_{n,k} - H(x)\right)^T \Sigma_{n,k}^{-1}\left(\hat{\theta}_{n,k} - H(x)\right)$$

$$p_n(t_k, x) = \frac{1}{C}\exp\left\{\phi_{n,k}\left(\hat{\theta}_{n,k}, H(x)\right)\right\} p_n^-(t_k, x) \tag{6.26}$$

Posterior target state estimate. For $n = 1, \ldots, N$

$$\hat{s}_{n,k} = \arg\max_{x \in S} p_n(t_k, x) \tag{6.27}$$

6.1.4 Variations

In the derivation of the MAP-PF algorithm in Sections 6.1.1 and 6.1.2 and the specific implementation given in Section 6.1.3, certain assumptions and choices are made to provide a straightforward presentation. Some of these may be modified to obtain variations on the MAP-PF algorithm, as discussed below.

We chose the quadratic penalty function in (6.11) for its simplicity and its relationship to the Gaussian probability density function. Other choices are possible. Any penalty function must be a continuous, differentiable function that is equal to zero when $\theta_k = \mathbf{H}(\mathbf{s}_k)$ and is negative otherwise. A key requirement for our development is that the penalty function be a sum of continuous, differentiable functions that each depend only on the state vector and auxiliary variable for a single target. Then the exponential of the penalty function behaves like the product of independent measurement likelihood functions, and the Bayesian filtering recursion decomposes into independent recursions for each target.

A key component of the MAP-PF algorithm is the extraction of measurements from the sensor data via the penalized maximum likelihood estimation problem in (6.14). It is a multidimensional optimization problem that requires joint maximization over the N components of θ_k. We chose to solve it by cyclically maximizing over each of the components while holding the others fixed. In [1], it was solved iteratively using the expectation maximization (EM) algorithm [9]. Suboptimal and/or approximate techniques may also be substituted here to reduce computational complexity. For example, we chose to perform only one iteration. In standard multiple target tracking techniques, contact measurements are obtained in a separate signal processing step that may be very simple or quite complicated. This signal processing step may be adapted into a "guided signal processing" step in which the signal processor is made aware of the number of targets and their estimated state to extract measurements for each target.

We have not specified a particular algorithm for the Bayesian filter recursions. In the next two sections, we provide a particle filter implementation and a linear-Gaussian implementation using the Kalman filter. However, any of the nonlinear filtering techniques discussed in Chapter 1 may be used.

To extract measurements from the sensor data, we required a point estimate of the target state vector $\hat{s}_{n,k}^-$ derived from the motion updated target state distribution $p_n^-(t_k, s_{n,k})$. For consistency with the MAP formulation, we chose it to be the maximum of the distribution, although this not required. We could choose it to be the mean, the median, or any other suitable estimate. For a Gaussian distribution, the mean and maximum are the same. For a unimodal distribution, the mean and maximum will be similar, while for a multimodal distribution, they may be very different. The choice of estimator will depend on the properties of the motion updated distribution and may also involve computational considerations.

6.2 PARTICLE FILTER IMPLEMENTATION

The particle filter implementation involves particle filter motion and information updates. We use the notation from Chapter 4 where the distribution for target n at time t_k is represented by a set of J_n particles and weights

$$\left\{\left(x_n^j(t_k), w_n^j(t_k)\right); j = 1, \ldots, J_n\right\}$$

An issue arises in the computation of the predicted and posterior target state estimates in (6.24) and (6.27). We cannot simply choose the particle with the largest weight before resampling because this will not generally coincide with the maximum of the distribution being represented. Furthermore, after resampling is performed, the particle weights at the motion update step are all equal to $1/J_n$, so there is no particle with the maximum weight. Finding the maximum requires computing a smoothed estimate of the distribution using a technique such as kernel density estimation and evaluating the maximum analytically. As discussed in Section 6.1.4, we may instead choose to use an alternative point estimate, such as the mean of the distribution. This is straightforward to implement for a particle filter, and will be the estimator used in this implementation.

Motion update. The motion update in (6.23) is given in Section 3.3.2. We obtain $x_n^j(t_k)$ by making an independent draw from the distribution

$$q_k\left(\cdot \mid x_n^j(t_{k-1}), n\right) \text{ for } j = 1, \ldots, J_n.$$

Predicted target state estimate. For the predicted state estimate in (6.24) we calculate the mean of the motion updated distribution

$$\hat{s}_{n,k}^- = \sum_{j=1}^{J_n} w_n^j(t_{k-1}) x_n^j(t_k) \tag{6.28}$$

Measurements. The measurements are obtained from (6.25).

Information update. The information update in (6.26) is performed by replacing the likelihood function in the information update step of the particle filter recursion in Section 3.2.2 with the exponential of the penalty function to obtain

$$w_n^j(t_k) = \frac{w_n^j(t_{k-1})\exp\left\{\phi_{n,k}\left(\hat{\theta}_{n,k}, H\left(x_n^j(t_k)\right)\right)\right\}}{\sum_{j'=1}^{J_n} w_n^{j'}(t_{k-1})\exp\left\{\phi_{n,k}\left(\hat{\theta}_{n,k}, H\left(x_n^{j'}(t_k)\right)\right)\right\}} \qquad (6.29)$$

Posterior target state estimate. We compute the posterior target state estimate in (6.27) as the mean of the posterior distribution

$$\hat{s}_{n,k} = \sum_{j=1}^{J_n} w_n^j(t_k)x_n^j(t_k) \qquad (6.30)$$

Resample. Finally, we resample the particles, as described in Section 3.3.3.

6.3 LINEAR-GAUSSIAN IMPLEMENTATION

Here we assume the linear-Gaussian motion model in (3.25). We also assume that the function in (6.7) that maps the target state to the auxiliary parameters is a linear function; that is,

$$\theta_{n,k} = H_{n,k}s_{n,k} \qquad (6.31)$$

This means that the parameter being measured by the sensor is a linear transformation of the target state. An example is when the target state is bearing and bearing rate and the sensor measurement is bearing.

   Substituting this into the quadratic penalty function in (6.11), the exponential of the quadratic function $\phi_{n,k}(\theta_{n,k}, H_{n,k}s_{n,k})$ is proportional to the Gaussian density $\eta(\cdot, H_{n,k}s_{n,k}, \Sigma_{n,k})$, and the motion and information updates take the form of the Kalman filter.

Motion update. The motion update in (6.23) is

$$\begin{aligned}
\mu_{n,k}^- &= F_{n,k}\mu_{n,k-1} + \varphi_{n,k} \\
P_{n,k}^- &= F_{n,k}P_{n,k}F_{n,k}^T + Q_{n,k} \\
p^-(t_k, x) &= \eta(x, \mu_{n,k}^-, P_{n,k}^-)
\end{aligned} \qquad (6.32)$$

Predicted target state estimate. For a Gaussian density the maximum and mean are the same, thus the predicted target state estimate in (6.24) is

$$\hat{s}_{n,k}^- = \mu_{n,k}^- \tag{6.33}$$

Measurements. The measurements are obtained from (6.25).

Information update. The information update in (6.26) is

$$
\begin{aligned}
K_{G_{n,k}} &= P_{n,k}^- H_{n,k}^T (H_{n,k} P_{n,k}^- H_{n,k}^T + \Sigma_{n,k})^{-1} \\
\mu_{n,k} &= \mu_{n,k}^- + K_{G_{n,k}} (\hat{\theta}_{n,k} - H_{n,k} \mu_{n,k}^-) \\
P_{n,k} &= P_{n,k}^- = (I - K_{G_{n,k}} H_{n,k}) P_{n,k}^- \\
p(t_k, x) &= \eta(x, \mu_{n,k}, P_{n,k})
\end{aligned}
\tag{6.34}
$$

Posterior target state estimate. The posterior target state estimate in (6.27) is the mean/maximum of the posterior distribution

$$\hat{s}_{n,k} = \mu_{n,k} \tag{6.35}$$

6.4 EXAMPLES

This section presents an example of the use of MAP-PF to determine bearing versus time tracks for acoustic targets from the sensor response of a linear array of omnidirectional hydrophones. We use the particle filter implementation of MAP-PF. We then compare the performance of MAP-PF to a particle filter version of the joint probabilistic data association (JPDA) algorithm given in Section 4.5.5.

6.4.1 Model

Figure 6.1 shows the target tracks in two-dimensional geographical coordinates for $N = 3$ targets observed for $K = 100$ scans by an $M = 20$-element uniformly spaced linear array with 1 m element spacing centered at $(0,0)$ and parallel to the horizontal axis in the figure. The time between scans is $\Delta t = 4$ seconds; thus the tracking interval is 400 seconds.

The target state x is the bearing, b, and bearing rate, $\dot{b}$, so that $x = (b, \dot{b})$. Here, we define the bearing as the angle measured counterclockwise from the horizontal array axis. This differs from the usual maritime convention where bearings are measured clockwise from north. Since there are three targets, the system state at time t_k is

$$\mathbf{s}_k = \left(\left(b_{1,k}, \dot{b}_{1,k}\right), \left(b_{2,k}, \dot{b}_{2,k}\right), \left(b_{3,k}, \dot{b}_{3,k}\right) \right)$$

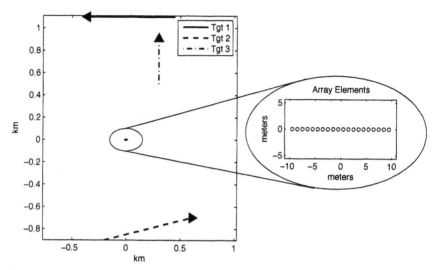

Figure 6.1 Three-target scenario in geographical coordinates. The array is a 20-element uniformly spaced linear array with 1m element spacing aligned along the horizontal axis and the tracking interval is 400s.

The bearing versus time plot for the three targets is shown in Figure 6.2. Note that bearings are only shown on the interval $b \in [0,180°)$. As discussed in Section 6.7, linear arrays have an ambiguity about the axis of the array, where a target with bearing $-b$ is indistinguishable from a target with bearing b. For this reason, we consider only the interval $b \in [0,180°)$. However, it must be understood that the actual bearing may be the "mirror bearing" in the interval $[-180°,0)$.

Although the targets are well separated in geographical space, the bearings seen by the array cross at two points during the track observation interval. Since target 2 is on the opposite side of the array from targets 1 and 3, the track crossings are due to the mirror ambiguity.

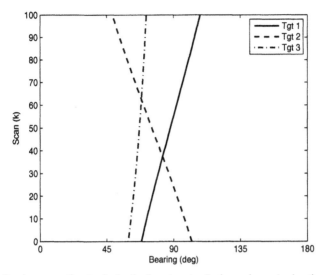

Figure 6.2 Bearing versus time tracks for the three targets. Each scan has a 4s duration.

6.4.1.1 Motion Model

We use the linear-Gaussian motion model 2 described in Section 3.2.1 with $\Delta_k = t_k - t_{k-1} = \Delta t = 4s$ and $w'_{n,k} \sim \mathcal{N}(0, Q')$, with $\sqrt{Q'} = 0.03$ deg/s. The initial target state and transition densities are

$$q_0(x \mid n) = \eta(x, \mu_{0,n}, P_{0,n}) \quad \text{and} \quad q_k(x \mid x', n) = \eta(x, Fx', Q) \tag{6.36}$$

where $x = (b, \dot{b})$ and

$$F = \begin{pmatrix} 1 & \Delta t \\ 0 & 1 \end{pmatrix} \quad \text{and} \quad Q = \begin{pmatrix} 0 & 0 \\ 0 & Q' \end{pmatrix} \tag{6.37}$$

6.4.1.2 Sensor Observation Model

The sensor array observation model is developed in Section 6.7 and summarized here. The observation vector y_k at time t_k is a 20×1 vector of complex hydrophone responses due to the $N = 3$ targets plus noise,

$$y_k = \sum_{n=1}^{3} v(b_{n,k}) \alpha_{n,k} + \varepsilon_k \tag{6.38}$$

The vector $v(b_{n,k})$ is the 20×1 "array response vector," which models the hydrophone responses for a target with bearing $b_{n,k}$. It has the form:

$$v(b_{n,k}) = \left(1, e^{j\pi \cos b_{n,k}}, e^{j2\pi \cos b_{n,k}}, \ldots, e^{j19\pi \cos b_{n,k}}\right)^T \quad (6.39)$$

The term $\alpha_{n,k}$ is the complex frequency response due to target n at time t_k at the first hydrophone. The term ε_k is a 20×1 vector of complex frequency responses due to the noise. We assume that it is a zero-mean complex Gaussian vector with covariance matrix $\sigma_\varepsilon^2 I$. We denote the signal-to-noise ratio (SNR) of the nth target as

$$SNR_{n,k} = \frac{E\left[|\alpha_{n,k}|^2\right]}{\sigma_\varepsilon^2} \quad (6.40)$$

We define b_k and α_k to be the 3×1 vectors of bearings and target responses,

$$b_k = \left(b_{1,k}, b_{2,k}, b_{3,k}\right)^T \text{ and } \alpha_k = \left(\alpha_{1,k}, \alpha_{2,k}, \alpha_{3,k}\right)^T$$

and $V(b_k)$ to be the 20×3 matrix of array response vectors,

$$V(b_k) = \left(v(b_{1,k}), v(b_{2,k}), v(b_{3,k})\right)$$

Then the observation vector in (6.38) may be written as

$$y_k = V(b_k)\alpha_k + \varepsilon_k \quad (6.41)$$

The log-likelihood function has the form

$$\ln L(y_k \mid b_k) = \frac{1}{\sigma_\varepsilon^2}\left|P_{V(b_k)} y_k\right|^2 \quad (6.42)$$

where the matrix $P_{V(b_k)}$ is a 20×20 projection matrix onto the three-dimensional subspace spanned by the columns of $V(b_k)$ and has the form:

$$P_{V(b_k)} = V(b_k)\left[V^{\#}(b_k)V(b_k)\right]^{-1} V^{\#}(b_k) \quad (6.43)$$

and # denotes the complex-conjugate (Hermitian) transpose.

6.4.2 MAP-PF Implementation

The natural measurement space of the sensor array is the target bearing; thus we choose the auxiliary parameters to be the bearings, which are related to the target state vector as follows

$$\theta_{n,k} = b_{n,k} = (1,0) \begin{pmatrix} b_{n,k} \\ \dot{b}_{n,k} \end{pmatrix} \tag{6.44}$$

Thus $H(s_{n,k}) = H s_{n,k}$ with $H = (1,0)$.

This model follows the linear-Gaussian model described in Section 6.3 and could be implemented with a Kalman filter; however, we will use the particle filter implementation described in Section 6.2 with $N_n = 1,000$ particles for each target. The particle filter implementation can easily be modified for nonlinear motion models, while the Kalman filter cannot.

We determine the penalty function matrix Σ_k by analyzing the CRB for jointly estimating the components of θ_k from the sensor data y_k. As shown in Section 6.7, the CRB $\mathbf{C}_k(\theta_k)$ is a 3×3 diagonal matrix whose *n*th component is:

$$[\mathbf{C}_k(\theta_k)]_{nn} = \left(2 SNR_{n,k} [B(\theta_k)]_{nn} \right)^{-1} \tag{6.45}$$

which is the CRB for estimating $\theta_{n,k}$ in the presence of the other two targets. The matrix $B(\theta_k)$ is defined in (6.71). It depends on the array geometry and the target angles and captures the interaction between targets in the joint estimation process. The CRB in (6.45) will be smaller for stronger targets and larger for weaker targets and will be larger when a target is close to another target. This provides a natural tightening or loosening of the penalty function for different target tracks and for different portions of the same target track.

Although the CRB in (6.45) is straightforward to calculate, it depends on the bearings and SNRs of the targets, which are not known. To calculate the CRB, we must either substitute presumed values or estimate them from the data. For the bearings, we select the bearing components of the predicted target state vectors $\theta_{n,k}^- = H \hat{s}_{n,k}^-$ and define $\hat{\theta}_k^- = \left(\hat{\theta}_{1,k}^-, \hat{\theta}_{2,k}^-, \hat{\theta}_{3,k}^- \right)$. The SNR is estimated from the data by first computing the ML estimate of α_k (derived in Section 6.7 and given in (6.59)) and then forming a smoothed estimate of SNR from

$$S\hat{N}R_{n,k} = \frac{1}{\sigma_\varepsilon^2} \left[\rho S\hat{N}R_{n,k-1} + (1-\rho) |\hat{\alpha}_{n,k}|^2 \right] \tag{6.46}$$

where ρ is a real value between 0 and 1 that controls the amount of smoothing. Then we can compute $\Sigma_{n,k}$ from

$$\Sigma_{n,k} = \left[\mathbf{C}_k(\hat{\boldsymbol{\theta}}_k^-)^{-1} \right]_{nn} = \left(2\, S\hat{N}R_{n,k} \left[B(\hat{\boldsymbol{\theta}}_k^-) \right]_{nn} \right)^{-1} \qquad (6.47)$$

The estimated CRB approach is adequate in this example; however, it can provide unreliable CRB values in other contexts. As discussed in Section 6.1.4, the remedy is to choose $\Sigma_{n,k}$ using the same rationale used for specifying the measurement covariance matrix in a standard Bayesian filter.

In Figure 6.3, we show the theoretical CRB and the estimated CRB for this example. We can see that the estimated CRB values are close to the theoretical CRBs. Target 3 has the highest SNR and the lowest CRB, while target 2 has the lowest SNR and highest CRB. During target crossings, the CRBs increase significantly. This occurs near scan k=37 for targets 1 and 2 and near scan k=62 for targets 2 and 3.

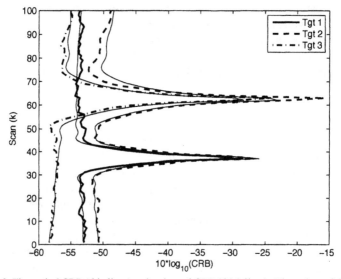

Figure 6.3 Theoretical CRB (thin lines) and estimated CRB (thick lines). The estimated CRB values are used in the tracker penalty function.

The bearing measurements are obtained by maximizing the penalized log-likelihood function

$$\hat{\theta}_{n,k} = \arg\max_{\theta} \left\{ \ln L_k(y_k \mid \hat{\theta}_{1,k}^-, \ldots, \hat{\theta}_{n-1,k}^-, \theta, \hat{\theta}_{n+1,k}^-, \ldots, \hat{\theta}_{N,k}^-) - \frac{1}{2\Sigma_{n,k}} \left(\theta - \hat{\theta}_{n,k}^- \right)^2 \right\} \quad (6.48)$$

where $\ln L(y_k \mid \boldsymbol{\theta}_k)$ is given in (6.42). For each target, the penalized log-likelihood function in (6.48) is a one-dimensional function obtained by holding the bearings of the other targets fixed. The first term in (6.48) is a one-dimensional "cut" of the multidimensional log-likelihood function.

Typical one-dimensional log-likelihood and penalized log-likelihood functions are shown in Figures 6.4 and 6.5 for each of the targets. Figure 6.4 shows the case where the targets are well separated at scan $k=19$. The log-likelihood functions for each of the targets have a peak near the true target location and suppressed peaks near the other target locations. Adding the penalty function further suppresses the "sidelobe" peaks and guides the processor to find the peak closest to the true target location to extract the bearing measurement.

Figure 6.5 shows the case where the targets 1 and 2 are close together in bearing at scan $k=35$. The log-likelihood function for target 3 again has a peak near the true target bearing and a suppressed peak near targets 1 and 2. Now the log-likelihood function for target 1 has a broad peak that is slightly suppressed near target 2's bearing, but the maximum value is shifted away from the true bearing of target 1. Adding the penalty function makes the peak narrower and pulls it back closer to the true bearing. The log-likelihood function for target 2 shows similar behavior. Note that the penalized log-likelihood functions for targets 1 and 2 are wider than in Figure 6.4, due to the CRB being larger.

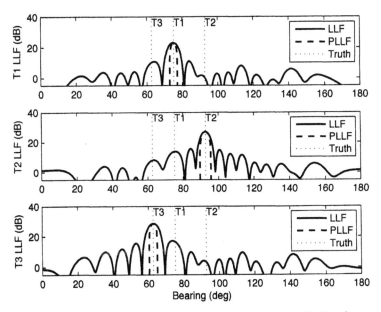

Figure 6.4 Log-likelihood and penalized log-likelihood functions for scan $k=19$ when the targets are well separated.

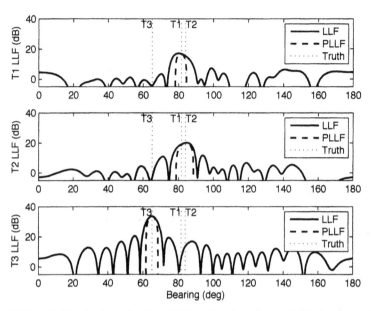

Figure 6.5 Log-likelihood and penalized log-likelihood functions for scan k=35 when targets 1 and 2 are close in bearing.

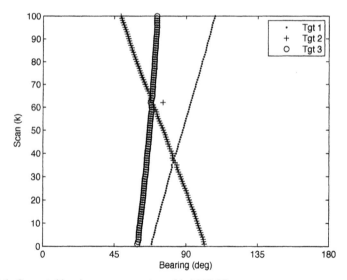

Figure 6.6 Generated bearing measurements used in MAP-PF.

Figure 6.6 shows the bearing measurements generated for each target. The bearing measurements are all close to the true target values, except for one measurement for target 2 at scan k=62. Because the tracker knows the number of targets present, one measurement is obtained for each target, and there are no missed detections. The penalty function acts like a prior distribution and guides the processing to find a peak in the likelihood function near the tracker-estimated target bearing. This simultaneously improves estimation performance and reduces false detections. Figure 6.7 shows the resulting bearing track estimates (on a different scale). The tracks are all close to the true tracks, even during target crossings. Note that the 2-σ error interval widens during target crossings due to the increase in the measurement variance as given by the CRB.

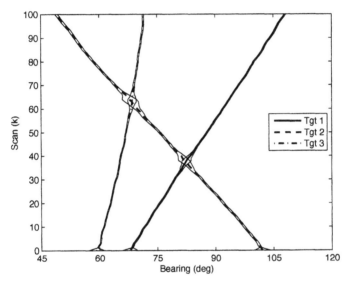

Figure 6.7 Bearing tracks (thick lines) and 2-σ error intervals (thin lines) for MAP-PF.

6.4.3 JPDA Implementation

We now demonstrate the behavior JPDA for the same example. We use the standard JPDA algorithm in which the number of targets is fixed and known to be three. The contacts are first obtained from the sensor data without knowledge of the number of targets. We use an iterative procedure in which we find the peak of the log-likelihood function as if there were only one target, then the peak of the log-likelihood function as if there were two targets with the first target known, and so on until the peak of the log-likelihood function falls below a threshold. The contact measurements obtained from this procedure are shown in Figure 6.8. For each scan, there are between two and five contact detections. For most scans there are three contact detections, with bearing measurements in the vicinity of the three target bearings. In some scans there are missed detections, and in some there are false alarms. Overall, the bearing measurements for JPDA show more variability than the measurements obtained by MAP-PF, which used prior knowledge of the number of targets and their locations from the current track estimates to aid in the measurement process.

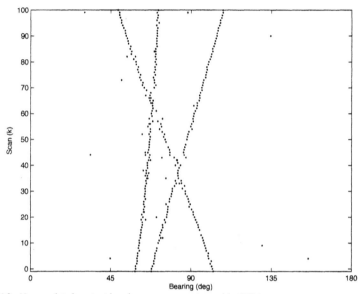

Figure 6.8 Unassociated contact bearing measurements used in JPDA.

The JPDA association procedure requires enumerating all possible association hypotheses and computing the association probabilities for the detected contacts. When there are three targets and two contacts, we indicate an association hypothesis by $\gamma = (n_1, n_2)$, where n_1 and n_2 are the targets to which contacts 1 and 2 are associated; $n_i = 0$ means that contact i is associated to a false alarm. For this case there are 13 association hypotheses as follows:

$$\gamma_1 = (0,0) \quad \gamma_2 = (0,1) \quad \gamma_3 = (0,2) \quad \gamma_4 = (0,3) \quad \gamma_5 = (1,0)$$
$$\gamma_6 = (2,0) \quad \gamma_7 = (3,0) \quad \gamma_8 = (1,2) \quad \gamma_9 = (1,3) \quad \gamma_{10} = (2,1) \qquad (6.49)$$
$$\gamma_{11} = (2,3) \quad \gamma_{12} = (3,1) \quad \gamma_{13} = (3,3)$$

For three targets and three, four, and five contacts, there are 34, 73, and 136 association hypotheses, respectively. We will not enumerate them here.

The motion model is the same as for MAP-PF. We use the scan association likelihood function model given in Section 4.5.2 with $f(y \mid n, x) = \eta(y, Hx, R)$ and parameters $p_d = 0.999$, $\rho = 0.1$, $V = 2$, $H = (1,0)$, and $R = 10^{-4} = -40$ dB. For JPDA, we use a constant measurement variance, R, that is larger than the CRB used in MAP-PF to account for the increased variability in the JPDA contact measurements. We use the particle filter implementation given in Section 4.5.5 with $N_n = 1,000$ particles for each target.

Figure 6.9 shows the resulting track estimates and the 2-σ error intervals for JPDA. The JPDA track estimates show more variability and have wider error intervals than for MAP-PF, and JPDA has more trouble during target crossings.

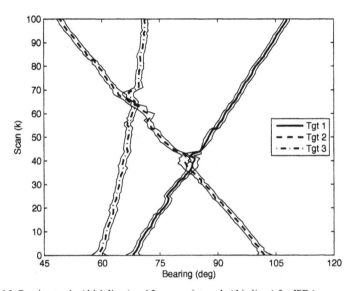

Figure 6.9 Bearing tracks (thick lines) and 2-σ error intervals (thin lines) for JPDA.

6.4.4 Summary of Examples

These examples have demonstrated that the MAP-PF approach to Bayesian multiple target tracking, where the observations are in the form unthresholded sensor data, can offer significant performance advantages over classical methods such as JPDA that start from contact measurements and must perform data association.

6.5 SUMMARY

This chapter develops the MAP-PF technique for Bayesian multiple target tracking under the assumption that we have access to unthresholded sensor data. The MAP-PF tracking algorithm performs both Bayesian filtering to obtain target state estimates and guided signal processing to obtain target measurements, whereas in classical methods the signal processing to generate contact measurements is performed separately from the tracking algorithm. In MAP-PF the two stages are coupled via the penalty function and the association step of classical methods is eliminated.

MAP-PF was formulated to find the MAP target state estimate from the posterior distribution at each scan. In order to facilitate the solution we introduced an auxiliary variable to represent a measurement in the natural measurement space of the sensor. This allowed us to use the penalty function method to decompose the problem into the solution of a sequence of simpler optimization problems. Estimating the auxiliary parameters turned out to be equivalent to generating measurements for each of the targets, while the quadratic penalty function had the interpretation of a set of independent measurement log-likelihood functions with the CRB characterizing the measurement error variance.

The solution to the MAP-PF optimization problem involved three levels of iteration. The outer level was the iteration in which the penalty weighting was made stronger at each iteration. The next level was the iteration between auxiliary parameter estimation and target state estimation. The third level was the iteration over the targets in the auxiliary parameter estimation step. In theory each iteration should be repeated until a convergence criterion is met, however multiple iterations at each level would make the implementation computationally infeasible in most applications. We provided a practical implementation that used only one iteration at each level.

In the derivation of the MAP-PF algorithm in Sections 6.1.1 and 6.1.2 and the specific implementation given in Section 6.1.3, certain assumptions and choices were made to provide a straightforward presentation. Modifications were discussed in Section 6.1.4, including penalty functions other than the quadratic penalty function, alternatives to penalized ML estimation for obtaining target measurements, and alternatives to using the maximum of the target state distribution as a point estimator. The Bayesian filter was left as a general recursion; however, specific implementations were given for the particle filter under a general class of models and for the Kalman filter under the linear-Gaussian model.

We provided an example comparing MAP-PF and JPDA for determining bearing versus time tracks for acoustic targets from the sensor responses of a linear array of omnidirectional hydrophones. The example demonstrated that the MAP-PF approach to Bayesian multiple target tracking, where the observations are in the form unthresholded sensor data, can offer significant performance advantages over classical methods that start from contact measurements and must perform data association.

Section 6.7 provides a description of the sensor array observation model and signal processing used in the examples in Section 6.4. The model is representative of a wider class of sensor models and Section 6.7 illustrates the methods and issues involved with obtaining measurements from the sensor data.

6.6 NOTES

MAP-PF was originally developed by Zarnich et al. [1] as a "batch" method in which the MAP point estimate of the target state was estimated jointly for all scans. A "sequential" version was given as an approximation to the batch method. The algorithm was derived for a narrowband sensor array model and a linear-Gaussian motion model, and the Bayesian filtering solution was the Kalman smoother.

Subsequent development of the method in [2–4] extended the sensor observation model to include multiple frequency bins and multiple sensor arrays, and extended the state vector to include the target spectral levels in each of the frequency bins. Including the spectrum in the state vector enables spectrally distinct targets to be tracked through a crossing without an increase in variance at the crossing that is so prominent in Figures 6.7 and 6.9. This is because spectral distinction makes targets identifiable. For the multiple sensor array model in [2, 3], the target state vector was the two- or three-dimensional position and velocity plus the spectral parameters, while the auxiliary parameters remained the bearings. The mapping from the target state to the auxiliary parameters was nonlinear and the Bayesian filtering solution was approximated by an extended Kalman filter.

In [5], the MAP-PF algorithm was applied to the over-the-horizon radar (OTHR) problem in which the natural measurements of the OTHR sensor are the azimuth angle, range, and Doppler frequency. Because of multipath propagation through the ionosphere, each target produced multiple detections or modes that had to be combined to form a single track in geographical coordinates.

The derivation given in this chapter started from the recursive Bayesian filtering model for multiple targets in Chapter 4 and is much more general than the derivations in [1–5] in terms of sensor and motion models. Consequently, we provided a direct derivation of the sequential MAP-PF method, in contrast to the previous derivations, which presented the sequential method as an approximation to the batch method. This formulation resulted in a more general algorithm that can be compared to classical techniques such as JPDA and allowed for the use of a wider class of Bayesian filtering techniques, including the particle filter.

6.7 SENSOR ARRAY OBSERVATION MODEL AND SIGNAL PROCESSING

This section provides a description of the sensor array observation model used in the examples in Section 6.4. We discuss the signal processing required to obtain "natural" parameter measurements as well as the CRB, which provides a bound on the measurement error variance. The model and the signal processing described here are representative of a wider class of sensor models and illustrate the methods and issues involved with obtaining measurements from the sensor data.

6.7.1 Sensor Observation Model

The sensor is an M-element uniformly spaced linear array centered at the origin and aligned along the horizontal axis, as shown in Figure 6.10.

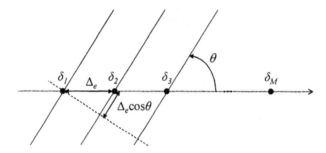

Figure 6.10 M-element uniformly spaced linear array aligned along the horizontal axis.

Let Δ_e denote the interelement spacing and δ_m denote the position on the horizontal axis of the mth element,

$$\delta_m = \left[m - \frac{M+1}{2} \right] \Delta_e; \qquad m = 1,\ldots,M \tag{6.50}$$

A propagating plane wave signal impinges on the array from angle θ, measured counterclockwise from the positive horizontal axis. The signal received at the mth element is a delayed version of the signal received at the first element, where the delay $\tau_m(\theta)$ at element m depends on the element position δ_m, the angle of arrival θ, and the plane wave's speed of propagation c as follows:

$$\tau_m(\theta) = -\frac{(\delta_m - \delta_1)\cos\theta}{c} = -(m-1)\frac{\Delta_e \cos\theta}{c} \tag{6.51}$$

We assume the sensor observations follow the "narrowband frequency-domain snapshot model" in Chapter 5 of [10]. In this model, an element response is produced for each scan by performing Fourier analysis on a segment of time series data to compute the complex Fourier coefficients of the frequency response in a discrete set of "frequency bins." The spectrum of the signal radiated by the target is assumed to be narrow with respect to the frequency bin width and only one Fourier coefficient has a signal contribution, while the rest contain only noise. The spectrum is assumed to be centered at frequency f_c, with corresponding wavelength $\lambda = c/f_c$. In the narrowband model, the complex frequency responses at each of the elements are the same except for a phase shift due to the signal delay. The $M \times 1$ array response vector $v(\theta)$ models the phase shifts at each of the elements for a planewave from angle θ,

$$v(\theta) = \left(1, e^{-j2\pi f_c \tau_2(\theta)}, e^{-j2\pi f_c \tau_3(\theta)}, \ldots, e^{-j2\pi f_c \tau_M(\theta)}\right)^T \tag{6.52}$$

If we make the standard assumption that the element spacing is half the signal wavelength (i.e., $\Delta_e = \lambda/2 = c/2f_c$), the array response vector simplifies to

$$v(\theta) = \left(1, e^{j\pi \cos\theta}, e^{j2\pi \cos\theta}, \ldots, e^{j(M-1)\pi \cos\theta}\right)^T \tag{6.53}$$

The array response vector in (6.53) is unique for $\theta \in [0, 180°)$; however, linear arrays have an ambiguity about the axis of the array. If we flip the diagram in Figure 6.10, we see that if a plane wave arrives from $-\theta$ the signal delays are the same as when the signal arrives from θ. The ambiguity is apparent in the array response vector, where $v(-\theta) = v(\theta)$ because $\cos(-\theta) = \cos(\theta)$. For this reason, we consider only the interval $\theta \in [0, 180°)$ for arrival angles; however, it must be understood that the actual arrival angle may be the "mirror angle" in the interval $[-180°, 0)$.

We assume that the observation vector y_k at time t_k is an $M \times 1$ vector of complex element responses due to N targets plus noise. Let $\alpha_{n,k}$ denote the complex frequency response at the first element due to the nth target at time t_k, $\theta_{n,k}$ denote the arrival angle of the nth target at time t_k, and ε_k denote the $M \times 1$ vector of complex frequency responses due to the noise. The observation vector has the form:

$$y_k = \sum_{n=1}^{N} v(\theta_{n,k}) \alpha_{n,k} + \varepsilon_k \tag{6.54}$$

We define $\boldsymbol{\theta}_k$ and $\boldsymbol{\alpha}_k$ to be the $N \times 1$ vectors of angles and target responses,

$$\mathbf{\theta}_k = \left(\theta_{1,k}, \theta_{2,k}, \cdots, \theta_{N,k}\right)^T \text{ and } \alpha_k = \left(\alpha_{1,k}, \alpha_{2,k}, \cdots, \alpha_{N,k}\right)^T$$

and $V(\mathbf{\theta}_k)$ to be the $M \times N$ matrix of array response vectors,

$$V(\mathbf{\theta}_k) = \left(v(\theta_{1,k}), v(\theta_{2,k}), \cdots, v(\theta_{N,k})\right)$$

Then the observation vector in (6.54) may be written as

$$y_k = V(\mathbf{\theta}_k)\alpha_k + \varepsilon_k \tag{6.55}$$

The noise vector ε_k is assumed to be a zero-mean complex Gaussian vector with covariance matrix $\sigma_\varepsilon^2 I$, and the target responses $\alpha_{n,k}$ are assumed to be uncorrelated, zero-mean complex random variables with variances $A_{n,k}$ that are uncorrelated with the noise. We denote the SNR of the nth target as

$$SNR_{n,k} = \frac{A_{n,k}}{\sigma_\varepsilon^2} \tag{6.56}$$

We follow the "conditional" model in Chapter 8 of [10] in which the likelihood function is a conditioned on α_k. For this model, the observation vector has a complex Gaussian distribution with mean $V(\mathbf{\theta}_k)\alpha_k$ and covariance matrix $\sigma_\varepsilon^2 I$; thus the conditional likelihood function has the form:

$$L(y_k \mid \mathbf{\theta}_k, \alpha_k) = \left(\pi\sigma_\varepsilon^2\right)^{-M} \exp\left\{-\frac{1}{\sigma_\varepsilon^2}\left(y_k - V(\mathbf{\theta}_k)\alpha_k\right)^\#\left(y_k - V(\mathbf{\theta}_k)\alpha_k\right)\right\} \tag{6.57}$$

where # denotes the complex-conjugate (Hermitian) transpose. The conditional likelihood function in (6.57) is our sensor array observation model. Sensor models of this form occur frequently in a variety of applications and are studied in detail in Chapter 5 of [8].

6.7.2 Array Signal Processing

We now turn our attention to the array signal processing required to obtain target angle measurements. Measurements are generated by estimating the parameter $\mathbf{\theta}_k$ from the sensor observations y_k. If we assume that $\mathbf{\theta}_k$ is a nonrandom parameter, we can find its maximum likelihood (ML) estimate by maximizing the likelihood function (or the log-likelihood function) with respect to $\mathbf{\theta}_k$. If we assume that $\mathbf{\theta}_k$ is a random variable with prior distribution $p(\mathbf{\theta}_k)$, we can compute the posterior distribution and find the MAP estimate by maximizing the posterior (or the logarithm of the posterior) with respect to $\mathbf{\theta}_k$. Other approaches are possible and many techniques have been proposed to solve this problem (e.g., see Chapter 9 of [10]). Here we focus on ML and MAP estimation.

First consider ML estimation. The conditional log-likelihood function has the form (dropping terms that do not depend on $\boldsymbol{\theta}_k$):

$$\ln L(y_k \mid \boldsymbol{\theta}_k, \alpha_k) = -\frac{1}{\sigma_\varepsilon^2}(y_k - V(\boldsymbol{\theta}_k)\alpha_k)^{\#}(y_k - V(\boldsymbol{\theta}_k)\alpha_k)$$

$$= -\frac{1}{\sigma_\varepsilon^2}|y_k - V(\boldsymbol{\theta}_k)\alpha_k|^2 \tag{6.58}$$

We cannot maximize this function with respect to $\boldsymbol{\theta}_k$ without knowledge of α_k, therefore we must maximize jointly with respect to both $\boldsymbol{\theta}_k$ and α_k. As shown in [8, 10], to maximize with respect to α_k, we take the (complex) derivative with respect to α_k, set the result equal to zero, and solve for α_k to obtain

$$\hat{\alpha}_k(\boldsymbol{\theta}_k) = \left[V^{\#}(\boldsymbol{\theta}_k)V(\boldsymbol{\theta}_k)\right]^{-1}V^{\#}(\boldsymbol{\theta}_k)y_k \tag{6.59}$$

This is the ML estimate of α_k when $\boldsymbol{\theta}_k$ is known. We now substitute this back into the log-likelihood function in (6.58) to obtain the "compressed" log-likelihood function

$$\ln L(y_k \mid \boldsymbol{\theta}_k) = \ln L(y_k \mid \boldsymbol{\theta}_k, \hat{\alpha}_k(\boldsymbol{\theta}_k)) = -\frac{1}{\sigma_\varepsilon^2}\left|\left[I - P_{V(\boldsymbol{\theta}_k)}\right]y_k\right|^2 \tag{6.60}$$

where

$$P_{V(\boldsymbol{\theta}_k)} = V(\boldsymbol{\theta}_k)\left[V^{\#}(\boldsymbol{\theta}_k)V(\boldsymbol{\theta}_k)\right]^{-1}V^{\#}(\boldsymbol{\theta}_k) \tag{6.61}$$

The matrix $P_{V(\boldsymbol{\theta}_k)}$ is an $M \times M$ projection matrix onto the N-dimensional subspace spanned by the columns of $V(\boldsymbol{\theta}_k)$. This subspace is called the "signal subspace" since the signal component of y_k in (6.55) is a linear combination of the columns of $V(\boldsymbol{\theta}_k)$ and is restricted to this subspace. The noise component ε_k may lie anywhere in the full M-dimensional observation space. The orthogonal projection matrix is defined as

$$P_{V(\boldsymbol{\theta}_k)}^{\perp} = I - P_{V(\boldsymbol{\theta}_k)} \tag{6.62}$$

This is an $M \times M$ projection matrix onto the $(M\text{-}N)$-dimensional subspace orthogonal to $V(\boldsymbol{\theta}_k)$.

When an $M \times N'$ matrix is premultiplied by an $M \times M$ projection matrix, the result is another $M \times N'$ matrix whose columns lie in the projection subspace, with any components from the orthogonal subspace removed. Some properties of the projection matrices in (6.61) and (6.62) are

(i) $P_{V(\theta_k)} V(\theta_k) = V(\theta_k)$

(ii) $P_{V(\theta_k)}^{\perp} V(\theta_k) = \mathbf{0}$

(iii) $P_{V(\theta_k)} P_{V(\theta_k)} = P_{V(\theta_k)}$ (6.63)

(iv) $P_{V(\theta_k)}^{\perp} P_{V(\theta_k)}^{\perp} = P_{V(\theta_k)}^{\perp}$

(v) $P_{V(\theta_k)}^{\perp} P_{V(\theta_k)} = \mathbf{0}$

where $\mathbf{0}$ denotes a matrix of zeros of the appropriate dimension.

The expression in (6.60) may be simplified by expanding the terms, using the properties in (6.63), and discarding terms that do not depend on θ_k. The result is:

$$\ln L(y_k \mid \theta_k) = \frac{1}{\sigma_\varepsilon^2} \left| P_{V(\theta_k)} y_k \right|^2 \qquad (6.64)$$

Thus the compressed log-likelihood function is proportional to the magnitude squared of the projection of the data onto the signal subspace. We will refer to $\ln L(y_k \mid \theta_k)$ in (6.64) as simply the log-likelihood function (LLF) going forward.

The ML estimate is found by maximizing the LLF with respect to θ_k. Ignoring the noise variance term, the ML estimation problem is:

$$\hat{\theta}_k = \arg\max_{\theta} \left| P_{V(\theta)} y_k \right|^2 \qquad (6.65)$$

Thus the ML estimator seeks to find the N-dimensional signal subspace that contains the most energy from the observed data. This problem does not have a closed form solution and must be evaluated using a numerical search procedure such as grid search, gradient search (e.g., Newton-Raphson), cyclic maximization (e.g., alternating projection), or iterative majorization (e.g., EM algorithm), as described in [6–11].

First consider the case of one target. Using the property that $v(\theta)^{\#} v(\theta) = M$, the one-dimensional ML estimation problem is

$$\hat{\theta}_k = \arg\max_{\theta} \left| \frac{v(\theta)^{\#} y_k}{M} \right|^2 \qquad (6.66)$$

The operation of multiplying y_k by $v(\theta)^\#/M$ is called conventional beamforming, and the conventional beam former (CBF) response $\left|v(\theta)^\# y_k/M\right|^2$ is obtained by "steering" the beamformer across all values of θ.

If we know that there is exactly one target present, then the ML estimate is obtained by finding the location of the largest peak in the CBF response. If we do not know whether a target is present or not, we can perform the same processing and declare a contact detection if the peak of the CBF response exceeds a threshold, with the corresponding contact angle measurement being the location of the peak.

Next consider the case of multiple targets. The ML estimation problem in (6.65) is a multidimensional optimization problem. While a simple grid search might be employed in the one- or two-dimensional cases, it quickly becomes infeasible in multiple dimensions, and an iterative procedure (gradient search, cyclic maximization, or iterative majorization) must be used. If we do not know the number of targets, we must first perform a model order estimation procedure. Several techniques are discussed in [10].

A suboptimal procedure in the multiple target case would be to compute the one-dimensional CBF response, which we would expect to have a peak whenever $v(\theta) = v(\theta_{n,k})$, and declare contact detections for all peaks that exceed the threshold, with the corresponding contact angle measurements being the locations of the peaks. This works well if the targets are well-separated and of similar strength. However, if targets are close together or one target is weaker than the other targets, the one-dimensional CBF procedure performs poorly.

The ML estimator is generally considered the "gold standard" processor for obtaining sensor measurements given that the number of targets is known. It is rarely implemented in practice because it is too computationally complex. The CBF is the most basic suboptimal processor and is widely used despite its limitations. Many techniques have been developed to provide improved performance over the baseline CBF processor while keeping complexity manageable. In any method, missed detections can arise due to insufficient target SNR, closely spaced targets, and masking of weak targets by strong targets. False detections can occur due to loud noise and spurious peaks (sidelobes) of the likelihood function due to strong targets.

We next consider MAP estimation, where we assume that θ_k is a random variable with prior distribution $p(\theta_k)$. We compute the posterior distribution and find the MAP estimate by maximizing the posterior (or the logarithm of the posterior) with respect to θ_k; that is,

$$\begin{aligned}
\hat{\theta}_k &= \arg\max_{\theta} \ln\left\{\frac{1}{C} L\left(y_k|\theta_k\right) p\left(\theta_k\right)\right\} \\
&= \arg\max_{\theta} \left\{\ln L\left(y_k|\theta_k\right) + \ln p\left(\theta_k\right)\right\}
\end{aligned} \tag{6.67}$$

Thus we add the logarithm of the prior distribution to the LLF and perform a multidimensional maximization as we did for ML estimation. In the MAP framework, the prior distribution includes knowledge of the number of targets, as well as information about the target angles. The MAP processor balances this information against the observed data to produce angle measurements.

6.7.3 Cramér-Rao Bound

Let $\Theta_k = (\theta_k, \alpha_k)$ denote the unknown parameters in the sensor array observation likelihood function in (6.57). The CRB is a matrix lower bound[1] on the covariance matrix of any unbiased estimator of Θ_k and is achievable asymptotically by the maximum likelihood estimator [8]. The CRB is obtained from the Fisher information matrix (FIM), defined as [8]:

$$\mathbf{J}_k(\Theta_k) = E_{y_k} \left\{ \left[\nabla_{\Theta_k} \ln L_k(y_k | \Theta_k) \right] \left[\nabla_{\Theta_k} \ln L_k(y_k | \Theta_k) \right]^T \right\}$$
$$= -E_{y_k} \left\{ \nabla_{\Theta_k} \left[\nabla_{\Theta_k} \ln L_k(y_k | \Theta_k) \right]^T \right\} \tag{6.68}$$

The inverse of the FIM is the joint CRB for estimating all components of Θ_k simultaneously

$$\mathbf{C}_k(\Theta_k) = \mathbf{J}_k(\Theta_k)^{-1} \tag{6.69}$$

The CRB for the nth parameter is obtained from the nth diagonal component (or nth diagonal block in the case of a vector parameter) of the joint CRB matrix.

The derivation of the CRB for jointly estimating θ_k and α_k is provided in Chapter 8 of [10]. The $N \times N$ block for θ_k has the form:

$$\mathbf{C}_k(\theta_k) = \frac{\sigma_\varepsilon^2}{2} \left(\mathrm{Re}\left[A_k \odot B(\theta_k) \right] \right)^{-1} \tag{6.70}$$

where $\odot$ denotes the Hadamard (element-wise) product, A_k is an $N \times N$ diagonal matrix with main diagonal entries $(A_{1,k}, \ldots, A_{N,k})$, and $B(\theta_k)$ is the $N \times N$ matrix

$$B(\theta_k) = D^{\#}(\theta_k) P_{V(\theta_k)}^{\perp} D(\theta_k) \tag{6.71}$$

where $D(\theta_k)$ is the $M \times N$ matrix of derivatives of the array response vectors,

---

[1] For real, symmetric matrices **A** and **B**, **A** is a lower bound on **B** in the sense that **B-A** is a positive semidefinite matrix. The diagonal entries in **A** are lower bounds on the diagonal entries in **B**.

$$D(\boldsymbol{\theta}_k) = \left(d(\theta_{1,k}), d(\theta_{2,k}), \cdots, d(\theta_{N,k}) \right)$$

$$d(\theta_{n,k}) = -j\pi \sin\theta_{n,k} \left(0, e^{j\pi \cos\theta_{n,k}}, 2e^{j2\pi \cos\theta_{n,k}}, \ldots, (M-1)e^{j(M-1)\pi \cos\theta_{n,k}} \right)^T \quad (6.72)$$

The matrix $B(\boldsymbol{\theta}_k)$ depends on the array geometry and the target angles and it captures the interaction between targets in the joint estimation process.

Since the off-diagonal entries in A_k are zeros, the off-diagonal entries in $A_k \odot B(\boldsymbol{\theta}_k)$ are also zeros. Furthermore since A_k is real and $B(\boldsymbol{\theta}_k)$ is Hermitian, the diagonal entries in $A_k \odot B(\boldsymbol{\theta}_k)$ are real. Therefore the CRB is a real diagonal matrix whose nth diagonal component is:

$$\left[C_k(\boldsymbol{\theta}_k) \right]_{nn} = \left(2\, SNR_{n,k} \left[B(\boldsymbol{\theta}_k) \right]_{nn} \right)^{-1} \quad (6.73)$$

which is the CRB for estimating $\theta_{n,k}$ in the presence of the remaining N-1 targets.

When there is one target, the CRB simplifies to:

$$C_k(\theta_k) = \frac{6}{SNR_k\, N(N^2-1)\pi^2 \sin^2\theta_k} \quad (6.74)$$

The CRB decreases as SNR increases and as the number of elements increases. It is inversely proportional to $\sin\theta_k$; therefore it is smallest when $\theta_k = 90°$ (array broadside) and increases as θ_k approaches $0°$ and $180°$ (array endfire).

It can be shown that the multiple target CRB in (6.73) is equal to the single target CRB in (6.74) plus a nonnegative term that is close to zero when the nth target is well separated from the other targets and increases rapidly as the nth target becomes close to any of the other targets. Thus the multiple target estimation problem has performance similar to the one target problem when the targets are well separated but suffers performance degradation when any two targets are close together. The longer the array, the closer the targets can be before degradation occurs.

References

[1]   Zarnich, R. E., K. L. Bell, and H. L. Van Trees, "A unified method for measurement and tracking of targets from an array of sensors," *IEEE Trans. on Signal Processing*, Vol. 49, No. 12, Dec. 2001, pp. 2950–2961.

[2]   Bell, K. L., "MAP-PF position tracking with a network of sensor arrays," *2005 IEEE Intl. Conf. on Acoustics, Speech, and Signal Processing*, Philadelphia, PA, March 2005, Vol. IV, pp. 849–852.

[3]   Bell, K. L., and R. Pitre, "MAP-PF 3D position tracking using multiple sensor arrays," *Fifth IEEE Sensor Array and Multichannel Signal Processing Workshop*, Darmstadt, Germany, July 2008, pp. 238–242.

[4] Bell, K. L., R. E. Zarnich, and R. Wasyk, "MAP-PF wideband multitarget and colored noise tracking," *2010 IEEE Intl. Conf. on Acoustics, Speech, and Signal Processing*, Dallas, TX, March 2010, pp. 2710–2713.

[5] Bell, K. L., "MAP-PF multi-mode tracking for over-the-horizon radar," *IEEE Radar Conference*, Atlanta, GA, May 2012, pp. 326–331.

[6] Nash, S. G., and A. Sofer, *Linear and Nonlinear Programming*, New York, NY: McGraw-Hill, 1996.

[7] Zangwill, W. I., *Nonlinear Programming: A Unified Approach*, Englewood Cliffs, NJ: Prentice-Hall, 1969.

[8] Van Trees, H. L., and K. L. Bell, *Detection, Estimation, and Modulation Theory, Part I (2nd Ed.)*, Hoboken, NJ: John Wiley & Sons, 2013.

[9] Green, P. J., "On use of the EM algorithm for penalized likelihood estimation," *Journal of the Royal Statistical Society, Series B*, Vol. 52, No. 3, 1990, pp. 443–452.

[10] Van Trees, H. L., *Optimum Array Processing: Detection, Estimation, and Modulation Theory, Part IV*, New York, NY: John Wiley and Sons, 2002.

[11] Stoica, P., and Y. Selen, "Cyclic minimizers, majorization techniques, and the expectation-maximization algorithm: a refresher," *IEEE Signal Processing Magazine*, Vol. 21, No. 1, Jan. 2004, pp. 112–114.

Chapter 7

Likelihood Ratio Detection and Tracking

This chapter explores the problem of detection and tracking when there is at most one target present. This problem is most pressing when signal-to-noise ratios (SNRs) are low or clutter rates are high. We might be performing surveillance of a region of the ocean's surface hoping to detect a periscope, if present, in the clutter of ocean waves. We might be scanning the horizon with an infrared sensor trying to detect a cruise missile at the earliest possible moment [1]. Both of these problems have two important features: (1) a target may or may not be present; (2) if a target is present, it will not produce a strong enough signal to be detected on a single glimpse by the sensor. As a cruise missile approaches, the SNR of the return will become large enough to call a detection on a single glimpse, but by then it may be too late to defend against the missile.

In active sonar tracking problems, the difficulty is that there are many returns that cross a threshold to provide detections, but most of these are clutter detections, not due to a target.

Likelihood ratio detection and tracking (LRDT) is based on an extension of the single target tracking methodology, presented in Chapter 3, to the case where there is at most one target present. However, the methodology also works well for certain multiple target detection and tracking problems as will be discussed below. In this chapter we present the theory and basic recursion for LRDT and show by examples how it can be successfully applied to low SNR and high clutter rate situations to produce high detection probabilities with low false alarm rates.

The chapter closes with a description of iLRT [2] which is a combination of multitarget intensity filtering presented in Chapter 5 and likelihood ratio tracking presented in this chapter. This combination provides a solution to two problems that confront intensity or probability hypothesis density (PHD) filtering, namely estimating the number of targets present and producing track estimates for the identified targets.

7.1 BASIC DEFINITIONS AND RELATIONS

We make the same basic assumptions as in Chapter 3. At the initial time $t = 0$ there exists a prior probabilistic description for the state of the target. This probability description is influenced in time by a set of K discrete observations or measurements $\mathbf{Y}(t) = (Y_1, ..., Y_K)$ obtained in the time interval $[0, t]$. The observations are received at the discrete (possibly random) times $(t_1, ..., t_K)$ where $0 < t_1 ... \leq t_K \leq t$. The measurements obtained at these various times need not be made with the same sensor or even with sensors of the same type. However, we do assume that, conditioned on the target's path, the statistics of the observations made at any time by any sensor are independent of those made at other times or by other sensors. There is an additional assumption concerning the target: We assume that the time evolution of its state is described by a Markov motion model.

The state space in which we detect and track targets depends upon the particular problem. Characteristically, the target state is described by a vector, some of whose components refer to the spatial location of the target, some to its velocity, and perhaps some to higher order properties such as acceleration. These components as well as others that may be important to the problem at hand can assume continuous values. Other elements that may be part of the state description may assume discrete values. Target class (type) and target configuration (such as periscope extended) are two examples.

An important practical consideration for a target detection and tracking problem is that we are not (usually) attempting to detect a target anywhere in the natural universe. Rather, we are determining only whether there is a target present in some surveillance region S. This region places bounds on various components of the target state vector. It typically bounds the spatial components in accordance with the surveillance limits of physical sensors and the velocity components in accordance with a priori limitations on the anticipated target velocities. These bounds are unremitting: If, for example, we process the data from a radar system designed to detect submarine periscopes, a target whose maximum speed is only a few knots, then the entry of a high-speed aircraft into the radar field of view might very well produce some confusing returns, but it could never be confused as representing a target traveling at submarine speeds. We also allow the possibility that no target exists in S. We shall formally refer to this state as the *null state* and designate it by the symbol ϕ.

We augment the target state space S with this null state to make $S^+ = S \cup \phi$. The augmented state space S^+ includes not only all of the vectors within S but the discrete null state ϕ as well. We shall assume there is a probability (density) function p defined on S^+. Since we are assuming that there is at most one target in S, we may write

$$p(\phi) + \int_S p(s)ds = 1$$

Both the state of the target $X(t) \in S^+$ as well as the information accumulated for estimation of the state probability densities evolve with time t. The process of target detection and tracking consists of computing the posterior version of the function p as new observations are available and propagating it to reflect the temporal evolution implied by target dynamics. Target dynamics include the probability of target motion into and out of the surveillance region as well as the probabilities of discrete target state changes.

Following the notation in Chapter 3, we let

$$p(t,s) = \Pr\left\{X(t) = s \mid \mathbf{Y}(t) = \left(Y(t_1), \ldots, Y(t_K)\right)\right\} \text{ for } s \in S^+$$

so that $p(t, \cdot)$ is the posterior distribution on $X(t)$ given $\mathbf{Y}(t)$. In this chapter, we shall assume that the conditions that insure the validity of the basic recursion for single target tracking in Chapter 3 hold so that $p(t, \cdot)$ can be computed in a recursive manner. Recall that

$$p^-(t_k, s_k) = \int_{S^+} q(s_k \mid s_{k-1}) p(t_{k-1}, s_{k-1}) \, ds_{k-1} \text{ for } s_k \in S^+$$

is the posterior from time t_{k-1} updated for target motion to time t_k, the time of the kth observation. Recall also that the definition of the likelihood function L_k. Specifically,

$$L_k(y_k \mid s) = \Pr\left\{Y_k = y_k \mid X(t_k) = s\right\} \tag{7.1}$$

where for each $s \in S^+$, $L_k(\cdot \mid s)$ is a probability (density) function on the measurement space M_k for the observation $Y_k = y_k$.

According to Bayes' Rule,

$$p(t_k, s) = \frac{p^-(t_k, s) L_k(y_k \mid s)}{C(k)} \qquad \text{for } s \in S$$

$$p(t_k, \phi) = \frac{p^-(t_k, \phi) L_k(y_k \mid \phi)}{C(k)} \tag{7.2}$$

In these equations the denominator is the probability of obtaining the measurement $Y_k = y_k$; that is,

$$C(k) = p^-(t_k, \phi)L_k(y_k \mid \phi) + \int_{s \in S} p^-(t_k, s)L_k(y_k \mid s)\,ds$$

7.1.1 Likelihood Ratio

We define the ratio of the state probability (density) to the null state probability $p(\phi)$ as the *likelihood ratio (density)* $\Lambda(s)$; that is,

$$\Lambda(s) = \frac{p(s)}{p(\phi)} \quad \text{for } s \in S \tag{7.3}$$

It would be more descriptive to call $\Lambda(s)$ the target likelihood ratio to distinguish it from the measurement likelihood ratio defined below. However, for simplicity, we will use the term likelihood ratio for $\Lambda(s)$. We will use notation that is consistent with that used for the probability densities. Thus, the prior and posterior forms of Λ become

$$\Lambda^-(t, s) = \frac{p^-(t, s)}{p^-(t, \phi)} \quad \text{and} \quad \Lambda(t, s) = \frac{p(t, s)}{p(t, \phi)} \quad \text{for } s \in S \text{ and } t \geq 0 \tag{7.4}$$

The likelihood ratio density has the same dimensions as the state probability density. Furthermore, from the likelihood ratio density one may easily recover the state probability density as well as the probability of the null state. Since

$$\int_S \Lambda(t, s)\,ds = \frac{1 - p(t, \phi)}{p(t, \phi)}$$

one easily finds that

$$p(t, s) = \frac{\Lambda(t, s)}{1 + \int_S \Lambda(t, s')ds'} \quad \text{for } s \in S$$

$$p(t, \phi) = \frac{1}{1 + \int_S \Lambda(t, s')ds'} \tag{7.5}$$

7.1.2 Measurement Likelihood Ratio

The measurement likelihood ratio $\mathcal{L}_k$ for the observation $Y_k = y$ is defined by

$$\mathcal{L}_k(y \mid s) = \frac{L_k(y \mid s)}{L_k(y \mid \phi)} \quad \text{for } y \in M_k, \ s \in S \tag{7.6}$$

$\mathcal{L}_k(y \mid s)$ is the ratio of the likelihood of receiving the observation $Y_k = y_k$ (given that the target is in state s) to the likelihood of receiving $Y_k = y_k$ given that no target is present. As discussed in [3], the measurement likelihood ratio has long been recognized as a prescription for optimal receiver design, and here we recognize that it plays large role in the overall process of sensor fusion.

7.2 LIKELIHOOD RATIO RECURSIONS

Under the assumptions for which the Bayesian recursion for single target tracking in Chapter 3 holds, we have the following extended recursion on the augmented state space S^+. After the final step of the recursion in Chapter 3, we have added the step of computing the posterior likelihood ratio function.

Extended Bayesian Recursion for Single Target Tracking

Initial distribution
$$p(t_0, s) = q_0(s) \ \text{ for } s \in S^+ \tag{7.7}$$

For $k \geq 1$ and $s \in S^+$,

Motion update
$$p^-(t_k, s) = \int_{S^+} q_k(s \mid s_{k-1}) p(t_{k-1}, s_{k-1}) ds_{k-1} \tag{7.8}$$

Measurement likelihood
$$L(y_k \mid s_k) = \Pr\{Y_k = y_k \mid X(t_k) = s_k\} \tag{7.9}$$

Information update
$$p(t_k, s) = \frac{1}{C} L_k(y_k \mid s) p^-(t_k, s) \tag{7.10}$$

Likelihood ratio
$$\Lambda(t_k, s) = \frac{p(t_k, s)}{p(t_k, \phi)} \tag{7.11}$$

The constant C in (7.10) is a normalizing factor that makes $p(t_k, \cdot)$ a probability (density) function. The above recursion can rewritten to be a recursion on the likelihood ratio function as follows. Write (7.8) as

$$p^-(t_k, s) = \int_S q_k(s \mid s_{k-1}) p(t_{k-1}, s_{k-1}) ds_{k-1} + q_k(s \mid \phi) p(t_{k-1}, \phi) \ \text{ for } s \in S$$

$$p^-(t_k, \phi) = \int_S q_k(\phi \mid s_{k-1}) p(t_{k-1}, s_{k-1}) ds_{k-1} + q_k(\phi \mid \phi) p(t_{k-1}, \phi) \tag{7.12}$$

From (7.4) it follows that

$$
\begin{aligned}
\Lambda^-\left(t_k,s\right) &= \frac{p^-(t_k,s)/p(t_{k-1},\phi)}{p^-(t_k,\phi)/p(t_{k-1},\phi)} \\
&= \frac{\int_S q_k(s\mid s_{k-1})\Lambda(t_{k-1},s_{k-1})ds_{k-1}+q_k(s\mid\phi)}{\int_S q_k(\phi\mid s_{k-1})\Lambda(t_{k-1},s_{k-1})ds_{k-1}+q_k(\phi\mid\phi)} \quad \text{for } s \in S
\end{aligned}
\tag{7.13}
$$

From (7.6) and (7.10) it follows that

$$
\Lambda(t_k,s) = \mathcal{L}_k\left(y_k,s\right)\frac{p^-(t_k,s)}{p^-(t_k,\phi)} = \mathcal{L}_k\left(y_k,s\right)\Lambda^-\left(t_k,s\right) \quad \text{for } s \in S
\tag{7.14}
$$

We can now state the likelihood ratio recursion.

Likelihood Ratio Recursion

Initial likelihood ratio

$$
\Lambda(0,s) = \frac{p(0,s)}{p(0,\phi)} \quad \text{for } s \in S
\tag{7.15}
$$

For $k \geq 1$ and $s \in S$

Motion update

$$
\Lambda^-(t_k,s) = \frac{\int_S q_k(s\mid s_{k-1})\Lambda(t_{k-1},s_{k-1})ds_{k-1}+q_k(s\mid\phi)}{\int_S q_k(\phi\mid s_{k-1})\Lambda(t_{k-1},s_{k-1})ds_{k-1}+q_k(\phi\mid\phi)}
\tag{7.16}
$$

Measurement likelihood ratio

$$
\mathcal{L}_k\left(y_k\mid s\right) = \frac{L\left(y_k\mid s\right)}{L\left(y_k\mid\phi\right)}
\tag{7.17}
$$

Information update

$$
\Lambda(t_k,s) = \mathcal{L}_k\left(y_k,s\right)\Lambda^-\left(t_k,s\right)
\tag{7.18}
$$

7.2.1 Simplified Likelihood Ratio Recursion

The recursion in (7.15) to (7.18) simplifies when probability mass flows from the state ϕ to S and from S to ϕ in such a fashion that the null state probability is not changed by the motion update; that is,

$$p^-(t_k,\phi) = q_k(\phi|\phi)p(t_{k-1},\phi) + \int_S q_k(\phi|s)p(t_{k-1},s)\,ds$$

$$= p(t_{k-1},\phi)$$

$$(7.19)$$

In this case the denominator in the top line of (7.13) equals 1, and we have that

$$\Lambda^-(t_k,s_k) = \frac{p^-(t_k,s)}{p(t_{k-1},\phi)} = \int_S q_k(s_k|s)\,\Lambda(t_{k-1},s)\,ds + q_k(s_k|\phi) \text{ for } s_k \in S \quad (7.20)$$

Now we can write a simplified version of the likelihood ratio recursion.

Simplified Likelihood Ratio Recursion

Initial likelihood ratio $\qquad \Lambda(t_0,s) = \dfrac{p(0,s)}{p(0,\phi)} \ \text{ for } s \in S \qquad (7.21)$

For $k \geq 1$ and $s \in S$,

Motion update $\qquad \Lambda^-(t_k,s) = \int_S q_k(s|s_{k-1})\Lambda(t_{k-1},s_{k-1})ds_{k-1} + q_k(s|\phi) \quad (7.22)$

Measurement likelihood ratio $\qquad \mathcal{L}_k(y_k|s) = \dfrac{L(y_k|s)}{L(y_k|\phi)} \qquad (7.23)$

Information update $\qquad \Lambda(t_k,s) = \mathcal{L}_k(y_k|s)\Lambda^-(t_k,s) \qquad (7.24)$

The simplified recursion is a reasonable approximation in problems where one is performing surveillance of a region that may or may not contain a target. Targets may enter and leave this region, but the probabilities of a target entering and leaving the region remain balanced.

As a special case, consider the situation where no mass moves from state ϕ to S or from S to ϕ under the motion assumptions. In this case $q_k(s|\phi) = 0$ for all $s \in S$ and $p^-(t_k,\phi) = p(t_{k-1},\phi)$ so that (7.22) becomes

$$\Lambda^-(t_k, s) = \int_S q_k(s \mid s_{k-1}) \Lambda(t_{k-1}, s_{k-1}) ds_{k-1} \qquad (7.25)$$

This is a very simple and convenient recursion, but its use can lead to problems because it does not allow for transitions from ϕ into S. Without such allowance it might be possible, after many observations indicating the absence of a target, that the prior likelihood ratio density would become so small for all states $s \in S$ that even very strong observational evidence for target presence would not raise the integrated likelihood ratio to a point where a target detection was called. The whole region S would have been deemed "sanitized" of target presence. By assumption, no targets are allowed to enter S so it would remain sanitized until strong evidence of a target presence was obtained from sensor observations. For most situations this is not realistic or desirable. Thus, providing for transitions from the null state to states within S is quite important to the proper functioning of the LRDT system discussed in this chapter.

Consider now (7.22) in the simplified recursion. For this recursion we have assumed that $p^-(t_k, \phi) = p(t_{k-1}, \phi)$. This means that the motion update step does not change the total probability associated with the null state. However, we do allow mass to move from ϕ to the states in S. We simply require that this movement must be matched by the same amount of mass moving from S to ϕ. This produces the motion update equation (7.22). The term $q_k(s \mid \phi)$ in (7.22) represents the motion from ϕ into S.

Usually one thinks of the term $q_k(s \mid \phi)$ as representing motion from ϕ into the boundary of S. However, the Markov motion model is capable of representing discontinuous transitions into and out of the space S other than those explainable in terms of normal kinematics of leaving or entering the boundary of S. It permits targets to suddenly materialize in the interior of S "out of thin air;" it also permits them to vanish just as unexpectedly. There are situations in which this does happen. Radar detection of a submarine periscope is an example. In effect, the target suddenly materializes when the periscope is raised. It remains present for a short period of time and then disappears. In this case the appearance and disappearance of the target does not have to take place at the boundary.

Even in the more usual situations where targets appear and disappear at the boundary, there is a practical reason for considering and employing discontinuous transitions in a model for state propagation. The motivation for introducing such artifacts into a problem involving physical targets like aircraft or missile launchers, none of which can actually appear or disappear unexplainably, is to make the system more robust to errors in the kinematic model.

7.2.2 Log-Likelihood Ratios

Frequently it is more convenient to write (7.24) in terms of the natural logarithms of the various quantities. One advantage of doing so is that it produces quantities that require less numerical range for their representation. Another advantage is that frequently the logarithm of the measurement likelihood ratio is a simpler function of the observations than is the actual measurement likelihood ratio itself. For example, when the measurement consists of an array of numbers, the measurement log-likelihood ratio often becomes a linear combination of those data, whereas the measurement likelihood ratio involves a product of powers of the data. In terms of logarithms, (7.18) becomes

$$\ln \Lambda(t_k,s) = \ln \Lambda^-(t_k,s) + \ln \mathcal{L}_k(y_k \mid s) \text{ for } s \in S \tag{7.26}$$

Because of the importance of the logarithm of the likelihood ratio density in this methodology, playing as large a role as the likelihood ratio density itself, we shall assign a special symbol to it. We define $\lambda = \ln \Lambda$ in all of its variations, employing λ^- to refer to the prior and leaving the unadorned symbol λ to refer to the posterior value. Equation (7.26) assumes the following form in this notation:

$$\lambda(t_k,s) = \lambda^-(t_k,s) + \ln \mathcal{L}_k(y_k \mid s) \text{ for } s \in S \tag{7.27}$$

Because the likelihood ratio density is not a scalar but has dimensions of the inverse of the state space volume, the quantity λ depends upon the units chosen to describe the state space of the target.

Measurement likelihood ratio functions are chosen for each sensor to reflect its salient properties including noise characterization and target effects. It is through these functions that we inject into the theory all that is required for making optimal Bayesian inferences from any measurements made with these sensors. Note that by the likelihood principle given in Chapter 2, the measurement likelihood function actually *defines* what is relevant to know about a sensor in order to perform this extraction of information.

7.3 DECLARING A TARGET PRESENT

The likelihood ratio methodology allows us to compute the Bayesian posterior probability density on the state space, including the discrete probability that no target resides in S at a given time. Thus it extracts all of the inferential content possible from the knowledge of the target dynamics, the a priori probability structure, and the evidence of the sensors. One may use this probability information any way one chooses in deciding whether a target is present or not.

Having said that, we shall present a number of traditional methods for making this decision. These methods are all based on the integrated likelihood ratio. Define

$$\overline{p}(t,1) = \int_S p(t,s)ds = \Pr\{\text{target present in } S \text{ at time } t\}$$

Then

$$\overline{\Lambda}(t) = \overline{p}(t,1) / p(t,\phi)$$

is defined to be the *integrated likelihood ratio at time t*. It is the ratio of the probability of the target being present in S to the probability of the target not being present in S at time t.

7.3.1 Minimizing Bayes' Risk

To calculate Bayes' risk, we must assign costs to the possible outcomes related to each decision, i.e., declaring a target present or not present. To do this, we define following costs:

- $C(1|1)$ if we declare the target present and it is present;
- $C(1|\phi)$ if we declare the target present and it is not present;
- $C(\phi|1)$ if we declare the target not present and it is present;
- $C(\phi|\phi)$ if we declare the target not present and it is not present.

We assume that it is always better to declare the correct state; that is,

$$C(1|1) < C(\phi|1) \text{ and } C(\phi|\phi) < C(1|\phi)$$

The *Bayes' risk* of a decision is defined to be the expected cost of making that decision. Specifically the Bayes' risk is defined as follows.

- $p(t,1)C(1|1) + p(t,\phi)C(1|\phi)$ for declaring a target present;
- $p(t,1)C(\phi|1) + p(t,\phi)C(\phi|\phi)$ for declaring a target not present.

One procedure for making a decision is to take that action that minimizes the Bayes' risk. Applying this criterion produces the following decision rule. Define the threshold

$$\Lambda_T = \frac{C(1|\phi) - C(\phi|\phi)}{C(\phi|1) - C(1|1)} \tag{7.28}$$

Then we declare the following.

- target present if $\overline{\Lambda}(t) > \Lambda_T$;
- target not present if $\overline{\Lambda}(t) \leq \Lambda_T$.

We see then that the integrated likelihood ratio is a sufficient decision statistic for taking an action to declare a target present or not when the criterion of performance is the minimization of the Bayes' risk.

7.3.2 Target Declaration at a Given Confidence Level

Another approach is to declare a target present whenever its probability exceeds a desired confidence level p_T. Again the integrated likelihood ratio is a sufficient decision statistic according to this criterion as well. The prescription is simply to declare a target present or not according to whether the integrated likelihood ratio exceeds a threshold, this time given by $\Lambda_T = p_T / (1 - p_T)$.

A special case of this is the *ideal receiver*, which is defined to be the decision rule that minimizes the average number of classification errors. Specifically, if we set

$$C(1|1) = 0, \ C(\phi|\phi) = 0, \ C(1|\phi) = 1, \text{ and } C(\phi|1) = 1$$

then minimizing Bayes' risk is equivalent to minimizing the expected number of miscalls of target present or not present. By (7.28) this is accomplished by setting $\Lambda_T = 1$. This corresponds to a confidence level of $p_T = 1/2$.

7.3.3 Neyman-Pearson Criterion for Declaration

Another standard approach in the design of target detectors is to declare targets present according to a rule that produces a specified false alarm rate. Naturally, the target detection probability must still be acceptable at that rate of false alarms. In the ideal case, one computes the distribution of the likelihood ratio with and without the target present and sets the threshold accordingly. Using the Neyman-Pearson approach, we find there is a threshold Λ_T such that calling a target present when the integrated likelihood ratio is above Λ_T produces the maximum probability of detection subject to the specified constraint on false alarm rate.

7.3.4 Track Before Detect

The process of LRDT is often referred to as *track before detect*. This terminology recognizes that we are performing tracking of a possible target (through computation of $p(t,\cdot)$) before we have decided to call a target present. The three examples given above show why the integrated log-likelihood ratio is so important for the track before detect process and provide a justification for devoting a chapter to LRDT even though it is already subsumed in the Bayesian tracking model of Chapter 3.

7.4 LOW-SNR EXAMPLES OF LRDT

This section presents two examples of LRDT in which the sensor responses are unthresholded and the SNR is low. The first example is a simplified one that illustrates the basic features of the methodology. This example shows how to integrate sensor responses on a moving target over time so that even though each response is well below the detection threshold, the log-likelihood ratio in the cell containing the target (which is changing over time) will build up to cross a detection threshold. This allows us to simultaneously declare a detection and estimate the target's state. This feature is particularly important in low SNR situations.

The second example gives more details of the periscope detection example presented in Chapter 1. In this example, we incorporate detailed physical models of the noise and signal into the measurement likelihood ratio. This enables us to achieve a high detection probability against a fleeting, low-SNR target in the presence of heavy clutter while maintaining a low false alarm rate.

7.4.1 Simple Example

For the first example, we present a simple situation that is not intended to be realistic but that is intended to illustrate the basic ideas of LRDT. The target state space is cellular with one dimension in position and one in velocity. At each time period, the sensor produces a measurement in each cell. The LRDT recursion uses these unthresholded responses to detect and track a simulated target with low SNR.

Target Motion Model

The target moves in a one-dimensional space of position cells represented by $\mathbf{J}$, the integers running from $-\infty$ to $+\infty$. The target can have one of nine base velocities where velocity is measured in cells per time step. The state space is

$$S = \mathbf{J} \times \{-3, -2, ..., 4, 5\}$$

so that a state $s \in S$ is an ordered pair (i, v) where i is position and v is base velocity. To S we adjoin the state ϕ, target not present, to obtain $S^+ = S \cup \{\phi\}$.

Let $X(t) = (i(t), v)$ (or ϕ) denote the target state at time t. The target motion model is given by

$$i(t+1) = i(t) + v + \delta(t) \tag{7.29}$$

where

$$\delta(t) = \begin{cases} 1 & \text{with probability } 0.1 \\ 0 & \text{with probability } 0.8 \\ -1 & \text{with probability } 0.1 \end{cases} \tag{7.30}$$

and $\delta(t)$ is independent of $\delta(t')$ for $t' \neq t$. Intuitively, the target has base velocity v, but there is noise in the velocity process so that the velocity varies randomly between $v-1$ and $v+1$. It is a discrete version of an almost constant velocity model. We assume that a target cannot transition from ϕ to S or vice versa.

For the initial distribution on target state, we assume that the probability that the target is not present is 0.75. If the target is present, then its position at time 0 has a uniform distribution on $\{1,...,100\}$, and its base velocity has a uniform distribution over the nine possible velocities. Specifically, the remaining 0.25 probability is spread uniformly over the 900 position-velocity pairs in $\{1,...,100\} \times \{-3, -2,...,4, 5\}$. The result is

$$p(0, \phi) = 0.75$$

$$p(0, (i, v)) = \frac{0.25}{900} = \frac{1}{3600} \quad \text{for } 1 \leq i \leq 100, \ -3 \leq v \leq 5$$

$$\lambda(0, (i, v)) = \ln \frac{p(0, (i, v))}{p(0, \phi)} = -\ln(2700) = -7.9$$

Sensor Model and Measurement Likelihood Function

The sensor for this example views only the spatial cells with indices between 1 and 100. At each time step, the sensor receives a normally distributed response $Y(i)$ from cell i for $1 \leq i \leq 100$. If the target is in cell j, for $1 \leq j \leq 100$, then $Y(j)$ has mean u and standard deviation 1. The target signal spills over to the adjacent cells so that the responses from those cells have mean $u/2$ and standard deviation 1. In all other cells, the response has mean 0 and standard deviation 1. Thus if the target is in cell j,

$$Y(i) \sim \begin{cases} \mathcal{N}(u,1) & \text{for } i = j \\ \mathcal{N}(\frac{1}{2}u,1) & \text{for } i = j-1 \text{ or } j+1 \\ \mathcal{N}(0,1) & \text{for } i < j-1 \text{ or } i > j+1 \end{cases} \tag{7.31}$$

Let $\mathbf{y} = (y_1, \ldots, y_{100})$ be the vector of responses obtained at a given time. Then the measurement likelihood function is

$$L(\mathbf{y} \mid (j,v)) = \left(\prod_{\substack{i < j-1 \\ i > j+1}} \eta(y_i, 0, 1) \right) \eta(y_{j-1}, \tfrac{1}{2}u, 1) \eta(y_j, u, 1) \eta(y_{j+1}, \tfrac{1}{2}u, 1) \text{ for } (j,v) \in S$$

$$L(\mathbf{y} \mid \phi) = \prod_{i=1}^{100} \eta(y_i, 0, 1)$$

where $\eta(\cdot, \mu, \sigma^2)$ is the density function for a $\mathcal{N}(\mu, \sigma^2)$ distribution. Note that the product in first line of the above equation omits factors with subscripts less than 1 or greater than 100. The likelihood function does not depend on velocity.

The measurement likelihood ratio is

$$\mathcal{L}(\mathbf{y} \mid (j,v)) = \begin{cases} \dfrac{\eta(y_j, u, 1)\eta(y_{j+1}, \tfrac{1}{2}u, 1)}{\prod_{i=j}^{j+1} \eta(y_i, 0, 1)} & j = 1 \\[3mm] \dfrac{\eta(y_{j-1}, \tfrac{1}{2}u, 1)\eta(y_j, u, 1)\eta(y_{j+1}, \tfrac{1}{2}u, 1)}{\prod_{i=j-1}^{j+1} \eta(y_i, 0, 1)} & 2 \leq j \leq 99 \\[3mm] \dfrac{\eta(y_{j-1}, \tfrac{1}{2}u, 1)\eta(y_j, u, 1)}{\prod_{i=j-1}^{j} \eta(y_i, 0, 1)} & j = 100 \end{cases} \tag{7.32}$$

The measurement log-likelihood ratio is

$$\ln \mathcal{L}(\mathbf{y} \mid (j,v)) = \begin{cases} uy_j + \dfrac{uy_{j+1}}{2} - \dfrac{5u^2}{8} & j = 1 \\[3mm] \dfrac{uy_{j-1}}{2} + uy_j + \dfrac{uy_{j+1}}{2} - \dfrac{3u^2}{4} & 2 \leq j \leq 99 \\[3mm] \dfrac{uy_{j-1}}{2} + uy_j - \dfrac{5u^2}{8} & j = 100 \end{cases} \tag{7.33}$$

Observe that the measurement likelihood ratio depends only on the measurement in the cell j and those on either side of it. That is, it depends only on the measurements in a local neighborhood of the cell j. This *local* property occurs when the likelihood function is a product of independent factors each of which depends only on the measurement in a single cell or a small neighborhood of cells. If the target signal was confined to a single cell in (7.31), then the measurement likelihood ratio would depend only the measurement in that cell.

Signal-to-Noise Ratio

To compute the SNR, we restrict our attention to the three cells containing the signal. In those cells we are dealing with a signal that has a three-dimensional Gaussian distribution with

$$\text{mean } \mu_1 = (u/2, u, u/2)^T \text{ and covariance } \Sigma = \mathbf{I}$$

where $\mathbf{I}$ is the three-dimensional identity matrix. The noise in those cells has a Gaussian distribution with mean $\mu_0 = (0,0,0)^T$ and covariance, $\mathbf{I}$. Thus the SNR is

$$\text{SNR} = \mu_1^T \Sigma^{-1} \mu_1 = \frac{3u^2}{2} \qquad (7.34)$$

7.4.1.1 Simulated Detection and Tracking Results

This section presents the results of applying LRDT to simulated data. To produce the data, we set the target signal level $u = 2$. This yields an SNR $= 6$ (7.8 dB).

The target starts at $i = 30$ at time 0 and moves with constant velocity $v = 1$. We used a simulation to produce sensor responses in the 100 cells viewed by the sensor for 16 time periods. Using these responses, we calculated the measurement log-likelihood ratio function in (7.33) for $t = 0,...,15$.

Since no probability mass moves from ϕ to S or from S to ϕ, we used the simplified likelihood ratio recursion with (7.25) in place of (7.22) to calculate the log-likelihood ratio function $\lambda(t,\cdot)$ for $t = 0,...,15$. To perform the step in (7.25), we exponentiated $\lambda(t_{k-1},s)$ to obtain $\Lambda(t_{k-1},s)$, performed the motion update to get $\Lambda^-(t_k,\cdot)$, and took logarithms to obtain $\lambda^-(t_k,s)$. Finally, we applied (7.27) to compute $\lambda(t_k,s)$. This is a simple but inefficient method of computing the likelihood ratio recursion. A more efficient method is described in Section 7.6.

Figures 7.1–7.4 show the log-likelihood ratio surface $\lambda(t,\cdot)$ at times $t = 0$, 5, 10, and 15, respectively. Note, that we show this surface only for the spatial cells with indices between 1 and 100 inclusive. The surface at time 0 (Figure 7.1) shows the effect of the observation at time 0. This surface is produced by adding the measurement log-likelihood ratio to the prior log-likelihood ratio. Since the prior log-likelihood ratio is a constant, equal to −7.9 for all states (except ϕ), the surface at time 0 is simply the measurement log-likelihood ratio for time 0 shifted down by 7.9 log-likelihood units. Since the likelihood function does not depend on velocity, the surface at time 0 shows no velocity dependence. The state of the target is shown by a vertical line dropped from a cross down to the log likelihood ratio surface at the position of the target's state, which is (30,1) at time 0. This surface is typical of the measurement log-likelihood ratio functions produced in the simulation. There are many peaks with roughly the same height as the one at the target state.

Figure 7.2 shows the log-likelihood ratio surface at $t = 5$. There is a peak developing in the vicinity of the target state at (35,1), but there are many noise peaks that are almost as high. By time $t = 10$, the picture (Figure 7.3) has clarified dramatically, the largest peak is clearly at the target state (40,1). The height of this peak is 29 log-likelihood units. There is still a substantial amount of background noise in the surface, although the peak levels are much below the one over the target state. At time $t = 15$ (Figure 7.4), the peak at the target state is equal to 31 log-likelihood units and most of the noise peaks have dropped below −30 log-likelihood units. Because of the dispersion in the velocity model (see (7.30)), the likelihood ratio at the target state is dispersed by each time step of motion. This produces the spread in the log-likelihood ratio peak at the target state that we see in Figures 7.3 and 7.4.

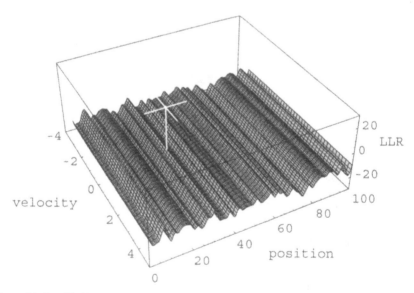

Figure 7.1 Log-likelihood ratio surface at $t = 0$. Position is given in cells and velocity in cells per time period.

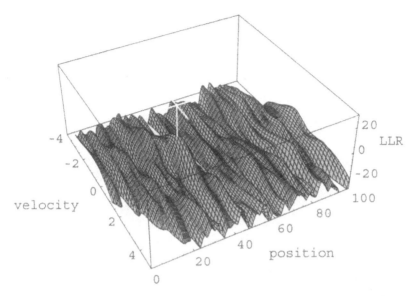

Figure 7.2 Log-likelihood ratio surface at $t = 5$. Position is given in cells and velocity in cells per time period.

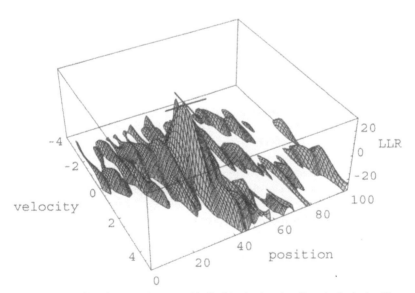

Figure 7.3 Log-likelihood ratio surface at $t = 10$. Position is given in cells and velocity in cells per time period.

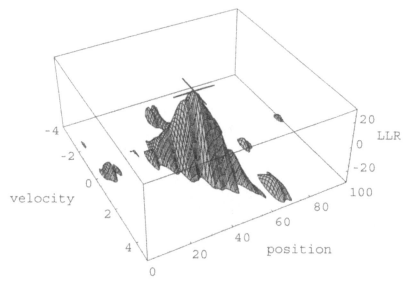

Figure 7.4 Log-likelihood ratio surface at $t = 15$. Position is given in cells and velocity in cells per time period.

7.4.1.2 Comparison to Matched Filter Detection

To call detections and estimate target state, we set a threshold λ_T and call a detection (target present) whenever the log-likelihood ratio surface rises above λ_T. The state at the peak of the surface can be used as a point estimate for the target state. In the case of Figure 7.4, we could provide a probability distribution for the target's state by fitting a bivariate normal distribution to the peak transformed to likelihood ratio units. This is done by fitting a quadratic to the peak in the log-likelihood ratio surface.

We now compare the effectiveness of LRDT to a more naïve detection scheme that is similar to matched filter processing. The matched filter operates by computing the dot product of the received measurement vector $\mathbf{y} = (y_1, \ldots, y_{100})$ in the 100 cells with the signal vector $\gamma^j = (\gamma_1^j, \ldots, \gamma_{100}^j)$ expected from a target in cell j for each cell. The matched filter output is

$$H(\mathbf{y} \mid j) = \sum_{j'=1}^{100} \gamma_{j'}^j y_{j'} \quad \text{for } j = 1, \ldots, 100$$

If the target is in cell j, then

$$\gamma_{j'}^{j} = \begin{cases} u & \text{if } j' = j \\ u/2 & \text{if } j' = j+1 \text{ or } j-1 \\ 0 & \text{otherwise} \end{cases}$$

with the obvious modification to account for the boundary cases when the cell is $j = 1$ or $j = 100$. Thus the matched filter output for a target in cell j is

$$H(\mathbf{y} \mid j) = \begin{cases} uy_j + \dfrac{uy_{j+1}}{2} & j = 1 \\ \dfrac{uy_{j-1}}{2} + uy_j + \dfrac{uy_{j+1}}{2} & 2 \le j \le 99 \\ \dfrac{uy_{j-1}}{2} + uy_j & j = 100 \end{cases} \tag{7.35}$$

As an alternate to computing the log-likelihood ratio surface at each time, we could simply compute the matched filter output for position j, for $j = 1, \ldots, 100$ from the sensor responses at each time and declare a detection (and target position estimate) whenever the response for some position j^* exceeds a specified threshold λ_M.

Section 6.2.2 of [4] compares the performance of this matched filter detection methodology, which does not integrate sensor responses over time, to that of LRDT. This comparison shows an impressive increase in performance produced from using LRDT. In particular, if one sets a detection threshold of $\lambda_T = 15$, then one obtains for LRDT a probability of detection $P_D(\lambda_T) \ge 0.93$ by time $t = 15$ with false alarm rate $\text{FAR}(\lambda_T) \le 7 \times 10^{-23}$ per spatial cell per time increment. To obtain the same detection performance from the matched filter detector, one has to choose a threshold of $\lambda_M = 1.04\sqrt{6} + 6$, which in turn produces a $\text{FAR}(\lambda_M) = 2.3 \times 10^{-4}$ per spatial cell per time increment.

To illustrate the difference in performance, let us scale the problem so that the sensor's view consists of 10,000 spatial cells and assume that there is a sensor response (or scan of) these cells every minute. In a 24-hour period this is 1,440 time increments. This will produce 3,312 false alarms per day for matched filter processing compared to 1.0×10^{-15} false alarms per day for LRDT; a stunning difference that illustrates the power of using LRDT for detecting low SNR targets.

7.4.2 Periscope Detection Example

Let us return to the periscope detection example given in Section 1.4. Recall that the radar generates returns in range and bearing having a 1-ft range resolution, 2-degree beamwidth, and a 5-Hz scan rate. It is assumed that a periscope will be exposed on the order of 10 seconds, although this information is not directly used by LRDT. Because the radar is mounted on a ship and is looking out to ranges of up to 10 nm, the grazing angle of the radar signal is low to the surface of the ocean. A high-resolution radar with a low grazing angle encounters significant clutter from breaking waves, which generate substantially higher returns than the mean ambient level. These high-intensity clutter spikes produce a high clutter rate and potentially a high false alarm rate; that is, a high rate of peaks in the radar return that "look like" targets.

7.4.2.1 Statistical Description of Radar Backscatter

A statistical characterization of the spiky clutter and the physical mechanisms have been the subject of both theoretical and experimental work [5–8]. One of the most successful approaches to describing the statistical behavior is a model with two time scales. In this two-scale model, the received signal is represented as a compound process where a fast speckle process is modulated by a slower process describing the scattering features in a radar cell. Over short time intervals (up to approximately 250 ms) the intensity observed in a fixed range cell is Rayleigh distributed. Over longer times, the mean of the Rayleigh component follows a Gamma distribution whose shape parameter is a function of the radar and ocean parameters such as grazing angle, resolution cell size, frequency, look direction, and sea state. The speckle depends on minute variations in the surface and is treated as uncorrelated noise. The underlying mean however depends on larger scale structures which can be modeled and tracked. Radar system noise is treated as being independent, additive, and complex Gaussian.

7.4.2.2 LRDT Description

To tackle the problem of detecting periscopes in heavy clutter, [9] developed an LRDT consisting of a clutter tracker and a target tracker. The clutter tracker estimates the mean clutter level in each range cell, every 1/5 of a second using the intensity of the radar returns.

The target tracker treats each beam of the radar separately. The nominal periscope is up for only a short period of time, so the chance of transiting from one beam to another is small, especially with overlapping beams. Within each beam, the target's state space is two-dimensional: *range* and *range rate*. Over the short period of exposure of a periscope, the tracker assumes a constant course and speed for the target. This produces a constant range rate for the target. Let $(r, \dot{r})$ represent a range and range-rate cell in the target state space. The likelihood ratio in cell $(r, \dot{r})$ is displaced to the cell $(r + \dot{r}\Delta t, \dot{r})$ over a time interval Δt. In this motion model, each velocity hypothesis can be treated independently.

The measurement used by the likelihood ratio tracker is the set of observed intensities of the radar return from each range cell. For each scan, the observed intensity and the estimate of the mean clutter level for a cell are used to compute a measurement likelihood ratio statistic in that cell. This statistic is the ratio of the probability of receiving the return given the target is in the cell (i.e., the periscope is exposed) to the probability of receiving the return given no target present. The two probabilities are conditioned on the estimate of the mean clutter level in the cell. This measurement likelihood ratio function is combined by pointwise multiplication with the motion-updated likelihood ratio surface for the target to produce the posterior likelihood ratio surface over the target state space. The resulting likelihood ratio surface is then updated for motion to form the motion-updated likelihood ratio surface for the next time increment.

Clutter Tracker—Estimating Mean Clutter

Let $y(t,x)$ be the radar intensity data at time t in range bin x. Recall that $y(t,x)$ is assumed to be the sum of an underlying mean clutter level denoted by $\gamma(t,x)$ and a zero-mean speckle term. We form an estimate of $\gamma(t,x)$ by using a weighted average of nearby cells using a Wiener filter as follows:

$$\hat{\gamma}(t,x) = \sum_{j=-M}^{N} \sum_{i=-A}^{B} h(i,j)y(t-j,x-i)$$

In the time dimension the sum extends over the past N scans, the current scan, and M scans into the future, while in the range direction it extends over the range bins from $x-B$ to $x+A$. The values of A, B, M, and N are determined by the correlation structure in the data. The filter coefficients $h(i,j)$ are determined from calculating the empirical covariance from a large sample of clutter data and can be updated in real time. The details of this calculation are given in Chapter 2 of [9].

A sketch of the data mask is shown in Figure 7.5. A variation of the clutter filter mask removes the data from the cell to be estimated in the linear filter. This is analogous to "guard cells" in cell-averaging constant false alarm rate detectors.

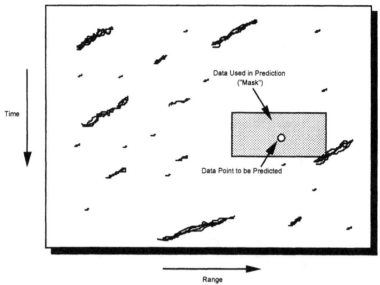

Figure 7.5 Clutter filter mask.

Measurement Likelihood Ratio Functions

The observation at time t is the received intensity $I(\cdot, t)$ over all radar range bins. The measurement likelihood ratio is a function of target position only since the radar is not measuring velocity. Let x be a hypothetical target position. Also, let

$$\gamma(t, x) = \text{mean clutter from state (cell) } x \text{ at time } t$$

In this section we shall assume that clutter is estimated with sufficient accuracy that we can take the estimate to be equal to the actual mean value. Reference [9] discusses the modification for uncertain clutter level.

Let $I(t, x)$ be the intensity received from cell x at time t and define

$$f_0\big(\zeta \mid \gamma(t, x)\big) = \Pr\big\{I(t, x) = \zeta \mid \text{mean clutter} = \gamma(t, x) \ \& \ \text{no target present at } x\big\}$$
$$f_1\big(\zeta \mid \gamma(t, x)\big) = \Pr\big\{I(t, x) = \zeta \mid \text{mean clutter} = \gamma(t, x) \ \& \ \text{target present at } x\big\}$$

Following the two-scale physical model for ocean clutter described above, we have that the distribution of the intensity $I(x, t)$ when no target is present is Rayleigh with mean $\gamma(x, t)$; that is,

$$f_0\big(\zeta \mid \gamma(t,x)\big) = \frac{1}{\gamma(t,x)}\exp\left(\frac{-\zeta}{\gamma(t,x)}\right) \text{ for } \zeta \geq 0 \qquad (7.36)$$

Fluctuating Radar Cross Section (RCS) Target

For this example, we assume the target has an exponentially fluctuating RCS. This is a Swerling I distribution [10] which describes a target whose RCS is constant during a scan but varies independently from scan to scan. In this case the response in intensity with target present is also exponential. Letting $w(t,x)$ be the mean intensity due to the target, we have

$$f_1\big(\zeta \mid \gamma(t,x)\big) = \frac{1}{\gamma(t,x)+w(t,x)}\exp\left(\frac{-\zeta}{\gamma(t,x)+w(t,x)}\right) \text{ for } \zeta \geq 0 \qquad (7.37)$$

The measurement likelihood ratio is thus

$$\mathcal{L}\big(I(t,x)\mid x,\gamma(t,x)\big) = \frac{f_1\big(I(t,x)\mid \gamma(t,x)\big)}{f_0\big(I(t,x)\mid \gamma(t,x)\big)}$$

$$= \frac{1}{1+w(t,x)/\gamma(t,x)}\exp\left(\left(\frac{I(t,x)}{\gamma(t,x)}\right)\frac{w(t,x)}{\gamma(t,x)+w(t,x)}\right) \qquad (7.38)$$

The LRDT recursion proceeds by using this expression, which is the likelihood ratio of observing the response $I(x,t)$ given the target is present at x to the likelihood of observing the response $I(x,t)$ given no target is present. Both the numerator and denominator of (7.38) depend on the mean clutter $\gamma(t,x)$ at x at time t.

Results

The above approach was applied to recorded radar data into which a target signal was injected as discussed in Section 1.4. Figures 1.14–1.16 show how the LRDT pulls a target out of radar returns in which no target is apparent. This provides an example of how a low-SNR target can be detected using unthresholded data by integrating measurement likelihood ratios of the sensor responses over time according to a target motion model. LRDT performs this integration in an efficient recursive manner. The measurement likelihood ratios are derived from detailed physical models of the radar return when no target is present and when a target is present.

7.5 THRESHOLDED DATA WITH HIGH CLUTTER RATE

LRDT can also be useful in cases where sensor data is thresholded to call detections but there is a high rate of clutter detections. Because LRDT integrates detection measurements over time, those sequences of detections that are consistent with the target motion model will be reinforced whereas most clutter detections will not. This allows LRDT to provide a high detection probability with a low false alarm rate.

To illustrate this, we consider a multistatic active sonar example involving sonobuoys. In an area of interest, there is a set of source and receiver sonobuoys whose locations are known. Typically there are more receiver than source buoys. Upon command, any one of the source buoys can send out a ping of acoustic energy. If there is a target submarine in the area, this energy will be reflected off the submarine and possibly produce a detection at one or more of the receiver buoys.

Typically, pings are activated from the various source buoys over a period time. Each ping produces set of detections and measurements at each receiver buoy. These measurements are sent to a central processing location with the goal of detecting and tracking a submarine if it is present.

This motivates the measurement likelihood functions and the example given in Section 7.5.2. While the example involves multi-static active sonar, the reader will see that the approach can be applied to many situations where there is a high rate of clutter contacts.

7.5.1 Measurement and False Alarm Model

At the time t_k (of the kth ping), there is a set $\mathbf{y}_m = \left(y_{m,1}, \ldots, y_{m,n(m)} \right)$ measurements generated by the mth sensor (receiver buoy). In the multistatic active case, the measurements are typically the bearing from the sensor (receiver buoy) to the echo and the time difference of arrival (TDOA) between the direct signal from the source buoy and the echo (reflected signal) from the target.

There are $M \geq 1$ sensors. Let L_m and $\mathcal{L}_m$ be measurement likelihood and the measurement likelihood ratio functions for the mth sensor. Then

$$\mathcal{L}_m \left(\mathbf{y}_m \mid s \right) = \frac{L_m \left(\mathbf{y}_m \mid s \right)}{L_m \left(\mathbf{y}_m \mid \phi \right)} \text{ for } s \in S \tag{7.39}$$

Let $\mathbf{y} = (\mathbf{y}_1, \ldots, \mathbf{y}_M)$. The composite measurement likelihood ratio for all sensors is given by

$$\mathcal{L}(\mathbf{y} \mid s) = \prod_{m=1}^{M} \mathcal{L}_m \left(\mathbf{y}_m \mid s \right) \tag{7.40}$$

In (7.40) we are assuming that the measurement likelihood ratios for the sensors are conditionally independent given target state.

7.5.1.1 Measurement Likelihood Ratio for Sensor m

We make the following definitions:

$p_d^m(s) =$ probability of sensor m detecting the target given it is in state s

$f_m(y \mid s) =$ probability (density) of obtaining measurement y given the target is detected at s

$\Phi_m =$ the number of false alarms received at sensor m

$$w_n(y_1, \ldots, y_n)$$
$$= \Pr\{\text{false alarm measurements} = (y_1, \ldots, y_n) \text{ at sensor } m \mid \Phi_m = n\} \quad (7.41)$$
$$= \prod_{i=1}^{n} w(y_i)$$

Using these definitions, we compute

$$L_m(\mathbf{y}_m \mid s) = p_d^m(s) \Pr\{\mathbf{y}_m \mid \text{target at } s \text{ and is detected}\} \\ + (1 - p_d^m(s)) \Pr\{\mathbf{y}_m \mid \text{target at } s \text{ and is not detected}\} \quad (7.42)$$

Let $(y_1, \ldots, \hat{y}_i, \ldots, y_n)$ denote the $n-1$ vector produced from $(y_1, \ldots, y_n)$ by deleting y_i. We assume that, given the target is detected, each of the measurements in $\mathbf{y}_m$ is equally likely to have been produced by the target. This yields

$$\Pr\{\mathbf{y}_m \mid \text{target is at } s \text{ and is detected}\}$$
$$= \frac{1}{n(m)} \sum_{i=1}^{n(m)} f_m(y_i \mid s) \Pr\{\Phi_m = n(m) - 1\} w_{n(m)-1}(y_1, \ldots, \hat{y}_i, \ldots, y_n) \quad (7.43)$$

Also,

$$\Pr\{\mathbf{y}_m \mid \text{target at } s \text{ and is not detected}\} = \Pr\{\Phi_m = n(m)\} w_n(y_1, \ldots, y_{n(m)}) \quad (7.44)$$

Substituting (7.43) and (7.44) into (7.42), we obtain

$$L_m \left(y_m \mid s \right) = \frac{p_d^m(s)}{n(m)} \sum_{i=1}^{n(m)} f_m \left(y_i \mid s \right) \Pr \left\{ \Phi_m = n(m) - 1 \right\} w_{n(m)-1} \left(y_1, \ldots, \hat{y}_i, \ldots, y_n \right)$$

$$+ \left(1 - p_d^m(s) \right) \Pr \left\{ \Phi_m = n(m) \right\} w_n \left(y_1, \ldots, y_{n(m)} \right) \tag{7.45}$$

For the target not present case, we compute

$$L_m \left(y_m \mid \phi \right) = \Pr \left\{ \Phi_m = n(m) \right\} w_n \left(y_1, \ldots, y_{n(m)} \right) \tag{7.46}$$

Taking the ratio of (7.45) and (7.46) and using (7.41), we obtain

$$\mathcal{L}_m \left(y_m \mid s \right) = \frac{p_d^m(s)}{n(m)} \sum_{i=1}^{n(m)} \left[\frac{\Pr \left\{ \Phi_m = n(m) - 1 \right\}}{\Pr \left\{ \Phi_m = n(m) \right\}} \right] \left(\frac{f(y_i \mid s)}{w(y_i)} \right) + \left(1 - p_d^m(s) \right) \tag{7.47}$$

7.5.1.2 Measurement Likelihood Ratio for Poisson-Distributed False Alarms

If we assume that the number of false alarms is Poisson-distributed with mean μ

$$\Pr \left\{ \Phi_m = n \right\} = \frac{\mu^n e^{-\mu}}{n!} \quad \text{for } n = 0, 1, \ldots$$

and (7.47) becomes

$$\mathcal{L}_m \left(\mathbf{y}_m \mid s \right) = p_d^m(s) \sum_{i=1}^{n(m)} \frac{f_m \left(y_i \mid s \right)}{\mu w(y_i)} + \left(1 - p_d^m(s) \right) \tag{7.48}$$

7.5.2 Multistatic Sonar Example

This section provides an example of the use of LRDT to detect and track targets with multistatic sonar in a high clutter rate situation. In this example there are two targets present in the region of interest, which is a 40 nm by 40 nm square. LRDT successfully detects and initiates a track on both targets using the grid-based implementation described in Section 7.6, which is based on the assumption that at most one target is present. In Section 7.7 we discuss why LRDT is able to detect and track separated multiple targets with no additional computational effort.

Figure 7.6 shows the 40 nm by 40 nm square region R of interest. There are nine sonobuoys located at the squares in Figure 7.6. Each buoy is a receiver, and the buoys at the corners can also act as sources for pings. The positions of the targets are marked by crosses. The targets are moving at a constant velocity. The target on the left is heading due north at 3 kn; the one on the right is heading due south at 6 kn. The target state space S is the set of vectors $s = (x, v)$ where $x \in R$ and v is a velocity whose speed is between 2 and 10 kn.

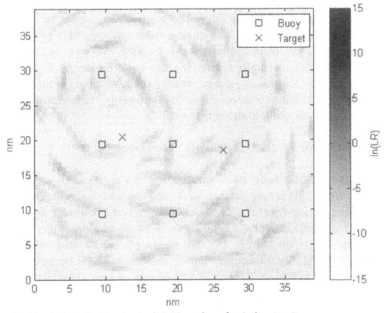

Figure 7.6 Marginal log likelihood ratio ln(LR) surface after 1 ping $(t = 0)$.

There is a ping every three minutes starting from the time of the first ping with a total of six pings. The pings start with the top left corner buoy and proceed cyclically through the bottom right, top right, and bottom left corner buoys. When a target is detected it produces a TDOA and bearing measurement. Both of these measurements have mean-zero Gaussian errors with a standard deviation of 0.5 s for the TDOA and 4° for the bearing. The receivers "listen" for a total 60 seconds after the source ping. However, no detections are reported from the m th sensor before the time of arrival $t_0(m)$ of the direct signal from the source. For each receiver, false detections are called at the times of a Poisson process with the rate $r = 0.5$ detections/s over the interval $[t_0(m), 60s]$. These times form the TDOAs for the false detections. When a false detection is called, the bearing is drawn from a uniform distribution over $[0°, 360°]$. Let $w_m(d, \theta)$ be the resulting density function for the false alarm measurements. These assumptions produced a approximately 175 false detections per ping for all nine sensors combined.

The location of the source buoy is known, so when a TDOA is measured at a receiver, it produces an ellipse of possible target locations with foci at the source and receiver buoy positions. The Gaussian error in the TDOA measurement "fattens out" this ellipse to represent the TDOA measurement likelihood function.

This function is multiplied by the bearing error measurement likelihood function resulting in combined likelihood function that is a "fat" arc of the TDOA ellipse.

Figure 7.6 shows the marginal log likelihood ratio surface after one ping. This surface is computed by summing the posterior likelihood ratios over all velocities at a spatial location and computing the natural logarithm of that sum to obtain the marginal log likelihood ratio at that location. Since we assume that the initial likelihood ratio is uniform over S, the posterior marginal at $t=0$ is proportional to the marginal log measurement likelihood ratio. In the figure one can see the "fat" elliptical arcs produced by the (mostly) false alarm detections and measurements resulting from the first ping.

Measurement Likelihood Ratio Function for Example

The measurement likelihood ratio function for this example is obtained from (7.40) and (7.48). Fix a ping (source buoy) and consider the mth receiver. For target position x let $t_m(x)$ be the TDOA at receiver m for the signal that is reflected off a target located at x, and let $b(x)$ be the bearing of the target from the receiver. Let d and θ be the measured TDOA and bearing. The measurement probability density function f_m in Section 7.5.1.1 becomes

$$f_m(d,\theta\,|\,x) = \eta\big(d,t_m(x),(0.5)^2 s^2\big)\eta\big(\theta,b(x),16\,(\deg)^2\big)$$

where $\eta(\cdot,\mu,\sigma^2)$ is the density function for a Gaussian distribution with mean μ and variance σ^2. For target detection probability we use

$$p_d^m(x) = \begin{cases} 0.25 & \text{if } t_m(x)\in[t_0(m),60s] \\ 0 & \text{otherwise} \end{cases} \tag{7.49}$$

Let $y_i = (d_i,\theta_i)$ for $i=1,\ldots,n(m)$ be the measurements produced by sensor m, and $\mathbf{y}_m = (y_1,\ldots,y_{m(n)})$. Then $\mathcal{L}_m$ in (7.48) becomes

$$\mathcal{L}_m(\mathbf{y}_m\,|\,s)$$
$$= p_d^m(x)\sum_{i=1}^{n(m)}\frac{f_m(d_i,\theta_i\,|\,x)}{r(60s-t_0)w_m(d_i,\theta_i)} + \big(1-p_d^m(x)\big) \text{ for } s=(x,v)\in S \tag{7.50}$$

The measurement likelihood ratio for the measurements produced by all receivers for a single ping is given by (7.40) with $M=9$.

LRDT Implementation

We used the grid-based implementation described in Section 7.6 to compute this example. The region R is represented by a 60 by 60 grid of square cells. There are 120 discrete velocities obtained from five speeds (2, 4, 6, 8, and 10 kn) and 24 courses spaced uniformly over $[0, 360°]$. We set the probability that the target is in S at time $t = 0$ equal to 0.5 so that $p(0,s) = 0.5/(60 \times 60 \times 120)$ for $s \in S$ and $p(0, \phi) = 0.5$. The initial likelihood ratio becomes

$$\Lambda(0,s) = \frac{0.5}{0.5(60 \times 60 \times 120)} \text{ for } s \in S$$

The target motion model used in LRDT assumes that the target speed is uniformly distributed between 2 kn and 10 kn and that velocity changes occur at exponentially distributed times with mean 1 hour. The distribution for a new velocity is the product of a uniform distribution over [2 kn, 10 kn] for speed and a uniform distribution over $[0°, 360°]$ for course.

Results

Figures 7.7–7.9 show the marginal log likelihood ratio surface after two, four, and six pings. One can see that LRDT has produced two peaks around the two targets while filtering out most of the false detections in spite of the 175 false detections per ping. The spread in the peaks is due to the motion model used by LRDT which allows for velocity changes. If we had used a constant velocity motion model, the peaks would have been sharper. For the simulation that produced the detections and measurements used in this example, we assumed a detection model that produces somewhat higher probabilities than are used in the measurement likelihood function. Our experience is that in practice LRDT performs well even with some mismatch between the detection and motion models used by LRDT and reality (the simulation in this case).

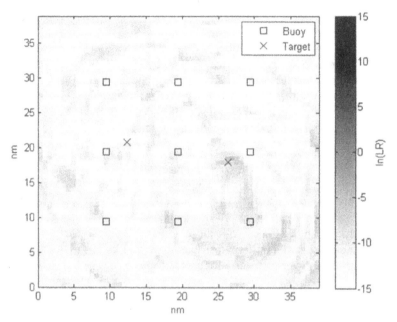

Figure 7.7 Marginal log likelihood ratio $\ln(\text{LR})$ surface after two pings $(t = 3$ minutes).

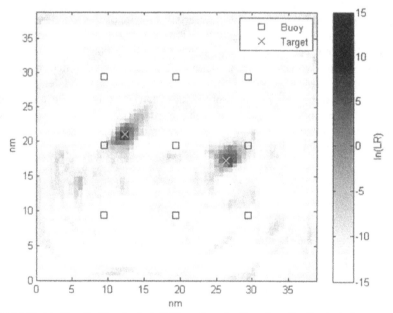

Figure 7.8 Marginal log likelihood ratio $\ln(\text{LR})$ surface after four pings $(t = 9$ minutes).

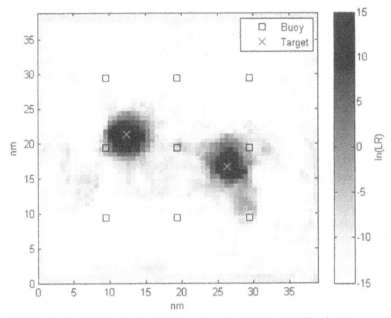

Figure 7.9 Marginal log likelihood ratio ln(LR) surface after six pings ($t = 15$ minutes).

7.6 GRID-BASED IMPLEMENTATION

This section describes a grid-based implementation of the simplified likelihood ratio recursion given in (7.21)–(7.24) for the case where the target state space is two-dimensional in position and two-dimensional in velocity.

Let $S = R \times V$ where R is a two-dimensional grid of J cells and $V = \{v_i : i = 1, \ldots, I\}$ is a discrete set of two-dimensional velocities. We represent S by I copies of R each with a velocity attached as shown in Figure 7.10. Each copy is called a *velocity sheet*.

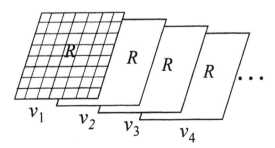

Figure 7.10 State space S for grid-based implementation of simplified likelihood ratio recursion.

7.6.1 Prior Likelihood Ratio

As in the example in Section 7.4.1, we specify a prior likelihood ratio for the each position-velocity cell at time 0. As an example if there are J position cells, then for the prior at $t = 0$ we could set

$$p(0,\phi) = 0.75$$

$$p(0,(j,v_i)) = \frac{0.25}{J \times I} \text{ for } 1 \le j \le J,\ 1 \le i \le I \qquad (7.51)$$

$$\Lambda(0,(j,v_i)) = \frac{p(0,(j,v_i))}{p(0,\phi)} = \frac{1}{3(J \times I)} \text{ for } 1 \le j \le J,\ 1 \le i \le I$$

7.6.2 Motion Model

We use a discrete time motion model with time increment Δ. This increment can be varied so that a motion update coincides with the time t_k of the kth measurement. One could also choose Δ to be small compared to the time between measurements, so that there will always be a motion update close to t_k.

At the end of a time increment, we assume

$$e^{-\alpha\Delta} = \Pr\{\text{no change in velocity}\} \text{ and } 1 - e^{-\alpha\Delta} = \Pr\{\text{velocity change}\} \quad (7.52)$$

for all states (j,v_i). If there is a velocity change, then the new velocity v_i' is drawn from the distribution

$$h(v) = \Pr\{v_i' = v\} \text{ for } v \in \{v_1,\ldots,v_I\} \qquad (7.53)$$

In (7.53) the distribution of the new velocity does not depend on the old velocity. One can allow this dependence if desired, but more computation is required. If we have limited knowledge of a target's possible velocities, we often take the new velocity distribution to be uniform over the velocity states.

We first describe the numerical implementation of this motion model for the special case where the target's velocity is constant (i.e., $\alpha = 0$). This is a good motion model to use when LRDT is being used to generate detections to hand off to a tracker in the manner described in Section 7.7.2.

7.6.2.1 Constant Velocity Motion

If the target velocity is constant, then the motion update is performed at each increment Δ by translating the velocity sheet for v_i by the vector displacement

Δv_i. This translation is performed for each velocity sheet. However, we want the spatial grids on the velocity sheets to stay in alignment, so the translation takes place only when the velocity sheet moves a full cell width in either direction. This is done as follows.

Let $(v_i(1), v_i(2))$ be the components of v_i in directions 1 and 2 (horizontal and vertical). For the k_Δth time increment and velocity v_i, we define counters $c_m(k_\Delta, v_i)$ for the grid displacement in the directions $m = 1, 2$. We initialize the counters as follows:

$$c_1(0, v_i) = c_2(0, v_i) = 0 \ \text{ for } i = 1, \ldots, I$$

Let d be the width of the square spatial cells. For the k_Δth motion update compute

$$c_m(k_\Delta, v_i) = c_m(k_\Delta - 1, v_i) + \Delta v_i(m) \tag{7.54}$$

If $\quad ld \leq c_m(k_\Delta, v_i) < (l+1)d \quad$ for some $\quad l = 1, 2, \ldots, \quad$ then we set $c_m(k, v_i) = c_m(k, v_i) - ld$ and translate the v_i velocity sheet l cells in the m direction. Otherwise we leave $c_m(k, v_i)$ as is and do not translate the velocity sheet in the mth direction. A velocity sheet can be translated in both directions during one motion update.

The result of doing this is that the velocity sheet for v_i moves at approximately the velocity v_i, and its spatial cells remain aligned with those of the original cells in R. As d becomes smaller, this approximation improves. When choosing the time increment Δ, one must take care that it is not so large that the grid cells are likely to translate more than one cell during an increment.

If a row or column of the velocity sheet moves outside the original grid for R, then that row or column is dropped from the velocity sheet and the row or column on the other side of the sheet that enters R is initialized with the prior likelihood ratio value given in (7.51).

7.6.2.2 Maneuvering Target

If the target can change velocities the motion update becomes more complicated. In this case we must add a "mixing" step after the translation to account for the possibility of a velocity change. Let

$\Lambda^*(t_k, (j, v_i)) =$ cumulative likelihood ratio in (j, v_i) after translation

Then the motion-updated cumulative likelihood ratio in (j, v_i) is computed by

$$\Lambda^-(t_k, (j, v_i)) = e^{-\alpha\Delta}\Lambda^*(t_k, (j, v_i)) + (1 - e^{-\alpha\Delta})h(v_i)\sum_{v'}\Lambda^*(t_k, (j, v')) \tag{7.55}$$

where the first term on the right-hand side is the product of the probability that there is no velocity change over the increment Δ times the likelihood ratio in that state (j, v_i). This is the amount of likelihood ratio that stays in (j, v_i). The second term accounts for likelihood ratio that moves out of some state (j, v') and into (j, v_i). Note that the sum on the right-hand side of (7.55) is the marginal likelihood ratio in cell j (i.e., the sum over all velocity sheets of the likelihood ratio in cell j). The product $(1 - e^{-\alpha \Delta}) h(v_i)$ equals the probability that there is a velocity change and that the new velocity is v_i. This is the fraction of likelihood ratio that moves out of other velocities and into v_i. This includes the possibility that there is a velocity change in state (j, v_i) but the new velocity chosen is v_i.

The translation step (7.54) accounts for motion among the spatial cells $j = 1, ..., J$. The mixing step (7.55) accounts for the transition from one velocity to another. If $\alpha = 0$ (constant velocity), then the motion update in (7.55) is the translation step.

7.6.3 Information Update

Suppose that we have received the measurements $\mathbf{y} = (\mathbf{y}_1, ..., \mathbf{y}_M)$ at time t_k and performed the motion update of the likelihood ratio velocity sheets to time t_k to obtain $\Lambda^-(t_k, (j, v_i))$. Let

$$\mathcal{L}(\mathbf{y} \mid (j, v_i)) = \prod_{m=1}^{M} \mathcal{L}_m(\mathbf{y}_m \mid (j, v_i)) \text{ for the state } s = (j, v_i)$$

Then

$$\Lambda(t_k, (j, v_i)) = \mathcal{L}(\mathbf{y} \mid (j, v_i)) \Lambda^-(t_k, (j, v_i)) \tag{7.56}$$

is the cumulative likelihood ratio in state (j, v_i) for $1 \leq j \leq J$, $1 \leq i \leq I$.

7.7 MULTIPLE TARGET TRACKING USING LRDT

In the discussion so far we have assumed that there is at most one target present in the area of interest. However, in many cases the measurement likelihood ratio has a local property that allows us to detect and track multiple targets. In this section we define and explore the consequences of this local property for multiple target tracking. The grid-based approach to implementing LRDT described above allows us to take advantage of this local property to perform multiple target detection and tracking with no additional computational cost.

7.7.1 Local Property for Measurement Likelihood Ratios

First we give some simple examples of the local property for measurement likelihood ratios. Suppose we are dealing with J discrete target states, $\{s_1,...,s_J\}$ corresponding to physical locations. Suppose that our observation is a vector $\mathbf{Y}$ that is formed from measurements corresponding to these spatial locations, so that $\mathbf{Y} = (Y(s_1),...,Y(s_J))$, where in the absence of a target in state s_j, the observation $Y(s_j)$ has a distribution with density function $\eta(\cdot,0,1)$ where $\eta(\cdot,\mu,\sigma^2)$ is the density function for a Gaussian distribution with mean μ and variance σ^2. The observations are independent of one another whether a target is present or not.

When a target is present in the jth state, the mean of $Y(s_j)$ is shifted from 0 to a value r. In order to perform a Bayesian update, we must compute the likelihood function for the observation $\mathbf{y} = (y(s_1),...,y(s_J))$ as follows:

$$L(\mathbf{y} \mid s_j) = \eta\big(y(s_j),r,1\big)\prod_{i \neq j}\eta\big(y(s_i),0,1\big)$$

$$= \exp\big(ry(s_j)-\tfrac{1}{2}r^2\big)\prod_{i=1}^{J}\eta\big(y(s_i),0,1\big)$$

By contrast the measurement likelihood ratio is

$$\mathcal{L}\big(\mathbf{y} \mid s_j\big) = \exp\big(ry(s_j)-\tfrac{1}{2}r^2\big) \text{ for } j=1,...,J \tag{7.57}$$

We see here a very characteristic result. Whereas the measurement likelihood function for the state s_j depends on the measurements from all states, the measurement likelihood ratio for s_j depends only on the data $y(s_j)$ from state s_j. This will be the case when the observation $\mathbf{Y}$ is a vector of independent observations as above

In some cases the measurement likelihood ratio for state s depends only on the data in a small neighborhood of s. The measurement likelihood ratio (7.32) for the state (j,v) in the simple example in Section 7.4.1 has this property. It depends only on the data from the three cells centered on position j. (It does not depend on the velocity v.)

The measurement likelihood ratio in (7.38) for the periscope detection problem depends only on the data from a single range and bearing cell given knowledge of the mean clutter level in that cell. Finally, the multistatic sonar measurement likelihood ratio function in (7.48) depends on only the detection probability for state s and measurement likelihood function evaluated at state s for the observed contacts. It does not depend on values associated with any state other than s.

We say that a measurement likelihood ratio is *localized* if the value for the state s depends only on the states in a "small" neighborhood of s. We call this the *likelihood neighborhood* of s.

The term small is not precisely defined here, but it means that the likelihood neighborhood is small enough that there is at most one target in it. If there are targets present at states s_1 and s_2 and the likelihood neighborhoods of these states do not overlap, then the likelihood ratio computations for these targets do not affect one another. If the response for one or both of the targets is strong enough, a peak will form independently of the other target.

Thus if two or more targets are present but well separated in the likelihood neighborhood sense, then a single likelihood ratio recursion can be performed and it will detect and (initiate) track on the targets independently. To emphasize, the likelihood ratio computation is the same whether there are 0, 1, or n targets in S so long as the necessary separation in likelihood neighborhoods prevails. Thus the grid-based implementation described above has a major advantage over particle filter implementations such as the one described in [11] which get increasingly complex as the number of possible targets increases.

7.7.2 LRDT as Detector for a Multiple Target Tracker

In multiple target situations where one is faced with low-SNR targets or high clutter rates, it is frequently valuable to use LRDT to provide detections to a multiple target tracker. LRDT is often able to maintain a high detection probability against targets while producing a low false alarm rate. This keeps the multiple target tracker from being overloaded by a large number of clutter or false alarm contacts. This generally improves the performance of the combined detector-tracker system compared to sending all contacts to the multiple target tracker.

An example is in multistatic active sonar systems. Usually there will be only one submarine target in the area of interest, but there are often multiple surface ships in the area that generate detections and tracks. In this case we use a constant velocity motion model in LRDT. This provides for optimum integration of the target signal over a short period of time. In particular, the likelihood ratio surface does not get diffused by the motion model. When a peak crosses a threshold, the likelihood ratio surface in the neighborhood of the peak is carved out and replaced by the prior likelihood ratio. The carved-out likelihood ratios are used to provide a contact with a probabilistic state estimate that is sent to a multiple target tracker. The multiple target tracker then decides whether to associate that contact with an existing track or form a new one. The tracker can be an MHT, a JPDA, or any other type of multiple target tracker. Generally the multiple target tracker will be designed to develop tracks over longer periods of time and to allow for target maneuvers. Thus LRDT acts as a short term detector-tracker. The multiple target tracker associates contacts to tracks and plays the role of a long-term tracker.

By returning the likelihood ratio values in the carve-out neighborhood to the prior value, we are restarting the LRDT in the neighborhood of the detection. Thus the next detection generated by the same target will have an independent measurement error. So long as the likelihood neighborhoods of the multiple targets in the area of interest do not overlap, the detection and carve out activities for one target will not affect the likelihood ratio build up for any other target. In practice there may be a small effect, but we have not found this to be a problem.

7.8 iLRT

This section presents a combination of intensity filtering and likelihood ratio tracking that overcomes one of the major difficulties with intensity filters, namely how to estimate the number of targets present and how to construct tracks from the sequence of intensity functions produced by intensity or PHD filters. The method for doing this is called iLRT; it was initially developed in [2].

Section 7.8.1 describes a particle filter implementation of the intensity filter recursion given by (5.28). The output of this particle filter is a set of paths whose density at each update time t_k approximates the intensity function produced by the recursion. In contrast to the usual particle filter, this one computes a separate cumulative likelihood ratio value for each path. It does this by initiating the cumulative value to be 1 at the beginning (birth) of the path. For each update time t_k after path initiation, the path's cumulative likelihood ratio value is multiplied by the value of the measurement likelihood ratio function for the measurements at time t_k evaluated at the state of the path at time t_k.

Section 7.8.2 describes how this cumulative likelihood ratio is used to call target detections and estimate target tracks. Section 7.8.3 applies iLRT to a multistatic active sonar example with multiple targets.

7.8.1 Particle Filter Implementation of Intensity Filtering

The intensity filter recursion is initiated by f_0 the intensity function (defined on the augmented state space S^+) at time $t_0 = 0$, which defines the Poisson point process (PPP) that provides the initial distribution on the number and location of targets in S^+. We assume

$$\mu_0 = \int_S f_0(s)ds < \infty$$

so that $f_0(s)/\mu_0$ for $s \in S$ is the probability density function for the state of a target in S. Recall that $f_0(\phi)$ is the expected number of clutter targets at time 0.

As in Chapter 5, we make the following assumptions. Measurements are received at a sequence of times $t_1, \ldots, t_K$. The target motion is Markovian with transition kernel $q_k(s_k \mid s_{k-1})$, which gives the probability that a target in state $s_{k-1} \in S^+$ at time t_{k-1} will transition to state $s_k \in S^+$ at time t_k. The same motion model applies to all targets. Define

$$P_k^d(s) = \text{probability of detecting a target located at } s \in S$$

$$P_k^d(\phi) = \text{probability of detecting a clutter target}$$

Given a target is detected at $s \in S^+$, the probability (density) that the measurement is equal to y is given by the likelihood function

$$l(y \mid s) = \Pr\{\text{measurement} = y \mid \text{target detected at } s\} \text{ for } s \in S$$

$$l(y \mid \phi) = \Pr\{\text{clutter measurement} = y \mid \text{clutter target detected}\}$$

The particle filter implementation contains two recursions. The first recursion is the intensity filter recursion given in Section 5.26, and the second is a cumulative likelihood ratio recursion.

7.8.1.1 Intensity Filter Recursion

Let f_{k-1} denote the posterior intensity function resulting from updating for motion and detections for times $t_1, \ldots, t_{k-1}$, and let f_k^- denote the motion updated intensity function for time t_k. The expected number of clutter detections at time t_k is $P_k^d(\phi) f_k^-(\phi)$. Each clutter detection produces a clutter measurement. Let $\mathbf{y}_k = (y_1, \ldots, y_{M_k})$ be the set of measurements at time t_k. Define

$$\overline{L}_k(\mathbf{y}_k \mid s) =$$

$$1 - P_k^d(s) + \sum_{r=1}^{M_k} \frac{l(y_r \mid s) P_k^d(s)}{l(y_r \mid \phi) P_k^d(\phi) f_k^-(\phi) + \int_S l(y_r \mid s') P_k^d(s') f_k^-(s') ds'} \tag{7.58}$$

Then the recursion in equation (5.28) may be written as

$$f_k(s) = f_k^-(s) \overline{L}_k(\mathbf{y}_k \mid s) \text{ for } s \in S \tag{7.59}$$

If $M_k = 0$, the summation on the right-hand side of (7.58) is taken equal to 0.

7.8.1.2 Measurement Likelihood Ratio

In order to compute the cumulative likelihood ratio on a particle path we must compute the measurement likelihood ratio $\mathcal{L}_k(\mathbf{y}_k \mid s)$ for the set $\mathbf{y}_k = (y_1,\ldots,y_{M_k})$ of measurements received at time t_k. Using a derivation similar to the one that produced the measurement likelihood ratio in (7.48) we obtain

$$\mathcal{L}_k(\mathbf{y}_k \mid s) = \left(1 - P_k^d(s)\right) + P_k^d(s) \sum_{r=1}^{M_k} \frac{l(y_r \mid s)}{l(y_r \mid \phi) P_k^d(\phi) f_k^-(\phi)} \tag{7.60}$$

7.8.1.3 Particles

To approximate the intensity function at time t_k we use N particles of the form

$$\left\{ x_j(t_k), w_j(t_k), \Lambda_j(t_k) \right\} \text{ for } j = 1,\ldots,N \tag{7.61}$$

where

$x_j(t_k)$ is a target path back to the time of origin $t_{o(j)}$ of the particle
$w_j(t_k) > 0$ is a weight
$\Lambda(t_k)$ is the cumulative likelihood ratio on the path $x_j(t_k)$

The particle filter approximates the density function f_k / μ_k at time t_k as follows. If $\mathcal{R}$ is a subset of S, then

$$\sum_{x_j(t_k) \in \mathcal{R}} w_j(t_k) \approx \int_{\mathcal{R}} \frac{f_k(s)}{\mu_k} ds \tag{7.62}$$

where the summation is taken over all particles whose position at time t_k is in $\mathcal{R}$. The cumulative likelihood ratio for $x_j(t_k)$ is

$$\Lambda_j(t_k) = \prod_{k'=o(j)}^{k} \mathcal{L}_{k'}\left(\mathbf{y}_{k'} \mid x_j(t_{k'})\right) \tag{7.63}$$

7.8.1.4 Particle Filter Recursion

Initialize the particle filter by making independent draws for the state $x_j(0)$ of particle j at time 0 from the probability density f_0 / μ_0 for $j = 1,\ldots,N$. Set

$$w_j(0) = 1/N \text{ and } \Lambda_j(0) = 1 \text{ for } j = 1,\ldots,N$$

Suppose that the recursion has produced

$$\left\{x_j(t_{k-1}), w_j(t_{k-1}), \Lambda_j(t_{k-1})\right\} \text{ for } j = 1, \ldots, N$$

Obtain $x_j(t_k)$ by making a draw from the Markov transition density $q_k(\cdot \mid x_j(t_{k-1}))$ and compute

$$
\begin{aligned}
w_j'(t_k) &= w_j(t_k)\overline{L}_k\left(\mathbf{y}_k \mid x_j(t_k)\right) \\
\Lambda_j(t_k) &= \Lambda_j(t_{k-1})\mathcal{L}_k\left(\mathbf{y}_k \mid x_j(t_k)\right) \text{ for } j = 1, \ldots, N
\end{aligned}
\tag{7.64}
$$

To account for the appearance of new targets (and to prevent particle starvation), we add N_0 new particles in S. They are distributed over S according to a "new target" distribution, which we take to be uniform over S. The particles drawn in this fashion are labeled $x_j(t_k)$ for $j = N+1, \ldots, N+N_0$. We specify $\nu_0 > 0$ and set the weight of the new particles to be

$$\tilde{w}_j(t_k) = \frac{\nu_0}{N_0}\overline{L}_k\left(\mathbf{y}_k \mid x_j(t_k)\right) \text{ for } j = N+1, \ldots, N+N_0 \tag{7.65}$$

and multiply the weights $w_j'(t_k)$ of the original particles by a constant factor to obtain $\tilde{w}_j(t_k)$ such that

$$\sum_{j=1}^{N+N_0} \tilde{w}_j(t_k) = N \tag{7.66}$$

We use $N = 20{,}000$, $N_0 = N/2 = 10{,}000$, and $\nu_0 = 0.01$ in the example below. The particles are resampled and perturbed using standard methods such as the ones described in Sections 3.3.3 and 3.3.4. During this process, split and perturbed particles retain their path history as discussed in Section 3.3.3, and the result is a set of N particles of the form given in (7.61) with weight $w_j(t_k) = 1/N$ for $j = j = 1, \ldots, N$.

7.8.2 Target Detection and Track Estimation

The cumulative likelihood ratio $\Lambda_j(t_k)$ is the likelihood ratio for all measurements received through t_k given that the target path is $x_j(t_k)$. The higher this ratio the more likely this target path is to represent a true target path. The target detection and track estimation procedure uses this statistic to determine the number of targets and their tracks as follows.

We set a threshold value τ. All particles with cumulative likelihood ratios greater than or equal to this threshold are called as detected. The detected paths are then clustered using the track heritage method. In this method, each particle is traced back to its origin. All paths having the same origin are clustered into a single target track. This identifies the number of targets present. The distribution for a target at time t_k is obtained from the states of the particles in the cluster forming that track. This distribution can be summarized by a mean and an approximating empirical covariance if desired.

An alternative method of estimating target state for a track is to use weights that are proportional to the cumulative likelihood ratio values for the particles that form a cluster. In the example given below this method produces better estimates of target state than those obtained from the weights $w_j(t_k)$ of the particles in the cluster.

The track heritage method works well for problems with high detection probability and relatively low clutter rate. However, performance degrades with increasing levels of clutter. (See [2].) Alternate clustering techniques such as agglomerative hierarchical clustering may produce better results.

7.8.3 Example

This example is taken from [2] and involves multistatic active sonar in a square region with sides of length 60 nm. The origin of the coordinate system is at the center of the square. Four sensors are located at the corners of a 15-km square centered at the origin. Their coordinates are $(\pm 7.5\,\text{km}, \pm 7.5\,\text{km})$ and their locations are shown as large black dots in Figures 7.11 and 7.12.

Figure 7.11 shows the results of 350 pings spaced 120 seconds apart. The target track is shown as a thick gray line. The target speed is 4 kn, approximately 2 m/s. On each ping, the probability of detection is a maximum of 0.3 if the acoustic signal hits and reflects from the target to the sensor at an angle near broadside and drops off as the angle from broadside increases. When the target is detected it produces a TDOA and bearing measurement. Both of these measurements have mean-zero Gaussian errors with standard deviations of 0.5s for the TDOA and 9° for the bearing. More details concerning this example are given in [2].

Over the 350 pings, there were 285 true detections and 976 false ones. Figure 7.11 shows that a single cluster or target track was formed and that it fits well with the actual target track. The thin grey lines are the paths of the particles in the cluster. The larger ellipse with the cross centered in it is the 3-σ ellipse corresponding the empirical covariance matrix computed from the locations and weights of the particles at the end of the 350 pings. The much smaller light-gray ellipse (see the inset) is the 3-σ ellipse computed using the normalized cumulative likelihood ratios of the particles as weights. This ellipse provides a better estimate of target position.

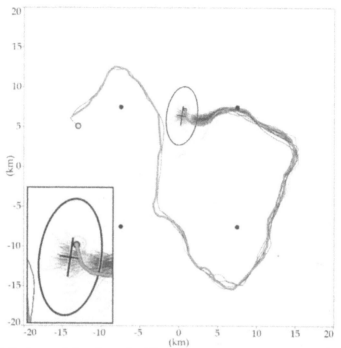

Figure 7.11 Single target in low clutter after 350 pings spaced 120s apart. The four sensors are located at the large black dots. The thick gray line is the actual track of the target. The particle paths in the cluster corresponding to a track are shown as thin grey lines. The larger ellipse is the 3-σ ellipse corresponding to covariance matrix computed from the particle weights. The smaller ellipse (see the inset) is computed using the scaled cumulative likelihood ratios of the particles for weights.

Figure 7.12 shows an example with two targets using target motion, detection, and clutter rate assumptions that are same as in the single target case above. Figure 7.12 shows the clusters after 325 pings spaced 120s apart. The thick grey lines are the two actual target tracks. The thin gray lines are the particle paths in the clusters. Again the larger ellipses are the 3-σ ellipse obtained from the particle weights and the smaller ones are the ellipses obtained from the normalized cumulative likelihood ratios. The iLRT has correctly determined that there are two targets present and formed good track estimates from the clusters corresponding to the targets.

This example shows that iLRT can successfully and accurately produce target tracks from the intensity functions computed by an intensity filter provided the clutter rate is low. Additional examples given in [2] show that the method has difficulty in high clutter rate situations (18,883 false detections in 350 pings) unless additional information such as feature information is included.

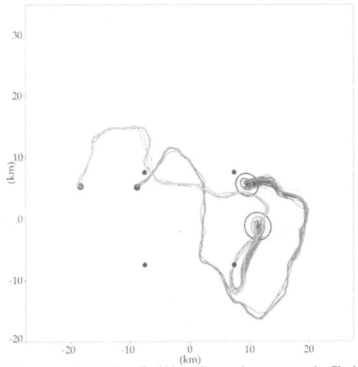

Figure 7.12 Two targets after 325 pings. The thick gray lines are the true target tracks. The thin grey lines are the particle paths. The darker lines correspond to one target the lighter ones to the other. For each target, the larger ellipse is the 3-σ ellipse computed from the particle weights, and the smaller ellipse is the one computed using the scaled cumulative likelihood ratios of the particles as weights. The sensor positions are given by the large black dots.

The inclusion of the cumulative likelihood ratio in the particle state is crucial for determining which paths are called as detections (target-like) and can therefore be clustered to estimate the number of targets present and their tracks. By raising or lowering the threshold τ for calling a path as detected, one can adjust the probability of detecting a target track. A lower value of τ produces a higher detection probability and higher false track rate. By varying τ, one can sweep out a receiver-operator curve of probability of target detection versus false track rate. Using this curve one can decide upon an operating point for the iLRT system. Typically one decides upon an acceptable false track rate and sets τ to achieve that rate. This will produce the highest detection probability with an acceptable false track rate.

7.9 SUMMARY

This chapter extends the single target tracking methodology developed in Chapter 3 to the case where there is either one or no target present. To do this we augment the state space S to $S^+ = S \cup \phi$ where ϕ represents no target present and define the likelihood ratio $\Lambda(s) = p(s) / p(\phi)$ for $s \in S$. We develop the likelihood ratio recursion that allows us to perform a Bayesian version of track before detect in which sensor data is incorporated recursively to compute posterior likelihood ratio surfaces. These surfaces can be used to determine whether a target is present and to estimate the state of a target if it is present. This is known as LRDT.

This chapter provides examples of applying LRDT to unthresholded sensor responses to detect and track targets in low-SNR situations where no single sensor response is strong enough to trigger a detection. LRDT accomplishes this by integrating the measurement likelihood ratio of sensor responses over time accounting for target motion. Sets of sensor responses that are consistent with the target motion model are reinforced to produce a peak at the target state. By contrast, responses due to noise generally do not reinforce one another. This allows LRDT to provide a high detection probability while maintaining a low false alarm rate. The measurement likelihood ratio can make use of detailed physical models as was done for the periscope detection example in Section 7.4.2 to further improve performance.

In situations where sensor responses produce detections with a high clutter rate, LRDT can also be applied to filter out clutter while maintaining a high detection probability. The multistatic active sonar example in Section 7.5.2 demonstrates that capability. As with the low-SNR examples, measurement likelihood ratios from sets of detections that are consistent with the target motion model will be reinforced to produce peaks in the posterior likelihood ratio surface whereas sets of clutter contacts will tend not to produce peaks. The example in Section 7.5.2 also shows that LRDT can be used to detect and initiate tracks on multiple separated targets with no increase in computational load and without having to estimate the number of targets present. Section 7.6 describes a grid-based implementation of LRDT. Section 7.7 discusses multiple target tracking using LRDT. The local nature of the measurement likelihood ratio function in many applications allows LRDT to do this for targets whose likelihood neighborhoods do not overlap (i.e., for separated targets). Section 7.7 also describes how LRDT can be used as a detector to feed a multiple target tracker in the case of low-SNR or high clutter rates. Using LRDT this way can provide a substantial reduction in the false detections sent to a multiple target tracker. It can also provide detections in low-SNR cases that require multiple sensor responses to produce a detection.

Section 7.8 presents iLRT, which is a combination of intensity filtering and likelihood ratio detection that overcomes the difficulty intensity or PHD filters have in determining the number of targets present and developing tracks for those targets.

7.10 NOTES

Use of the augmented state space S^+, which includes the state ϕ to represent target not present, to allow for both detection and tracking in the Bayesian recursion appears in a number of papers including [12–14].

Likelihood ratio tracking has been used in the defense community for a number of years, but Lindgren and Taylor [15] were the first (to our knowledge) to publish the method, which they called Bayesian field tracking, in the open literature. They derive the likelihood ratio recursion in (7.15) to (7.18) and propose a grid-based implementation different from the one in Section 7.6. They observe that the local property of the measurement likelihood ratio allows the method to be used for multiple target detection and track initiation with only minor changes to their algorithm. In [1] this methodology is applied to detection of low-SNR targets using an infrared sensor. Particle filter implementations of LRDT are described in [11, 16, 17]. Reference [11] discusses the extension of the particle filter approach to perform multiple target detection and tracking but notes that it becomes very complicated for more than three targets.

Traditional track-before-detect approaches hypothesize a large number of possible target tracks over a fixed time window and try to determine if there are subsets of the detections or sensor responses that are consistent with one or more of the tracks. Typically, this approach is applied to thresholded sensor data in the form of contacts (detections with measurements) and constant velocity tracks. For each possible track, this approach determines which contacts are associated with the track. If the window size is n scans of contacts, then a typical approach is to require $m \leq n$ associated contacts in the window before calling a detection. Hence the name track-before-detect.

This approach becomes computationally very expensive as the number of potential tracks increases. Imagine doing this for 100,000 potential tracks. By contrast LRDT's grid-based, recursive implementation considers a very large number of potential tracks with modest computational effort. As an example consider a standard LRDT implementation in which the grid consists of 200 by 200 spatial cells and 200 velocities. For a constant velocity model, this yields $(200)^3 = 8$ million potential tracks considered. For a maneuvering target model, this explodes to $(200)^2 \times (200)^n$ tracks for an n-step time window. Note that LRDT does not require the specification of a time window.

References

[1]   Merlo, S. P., R. G. Lindgren, and P. J. Davis, "Application of Bayesian field track-before-detect and peak likelihood track-after-detect trackers to shipboard infrared search and track," *Proceedings SPIE* Vol 3163, Signal and Data Processing of Small Targets, 1997, pp. 194-203.

[2]   Streit, R. S., B. Osborn, and K. Orlov, "Hybrid intensity and likelihood ratio tracking (iLRT) for multitarget detection," in *Proceedings 14th International Conference on Information Fusion*, Chicago, IL, USA, July 5-8, 2011.

[3]   Van Trees, H. L., and K. L. Bell, *Detection Estimation and Modulation Theory: Part 1 – Detection, Estimation, and Filtering Theory*, 2nd Ed., Hoboken NJ: John Wiley, 2013.

[4]   Stone, L. D., C. A. Barlow, and T. L. Corwin, *Bayesian Multiple Target Tracking*, 1st Ed, Norwood MA: Artech House, 1999.

[5]   Jakeman, E., and P. N. Pusey, "A model for non-Rayleigh sea echo," *IEEE Transactions on Antennas and Propagation*, Vol. AP-24, No. 6, 1977, pp. 806-814,

[6]   Ward, K. D., C. J. Baker, and S. Watts, "Maritime surveillance radar. Part 1: Radar scattering from the ocean surface," *IEE Proceedings*, Vol. 137, Pt. F, No. 2, 1990, pp.51-62.

[7]   Lee, P. H. Y., et al, "X band microwave backscattering from ocean waves," *Journal of Geophysical Research*, Vol. 100, No. C2, 1995, pp. 2591-2611.

[8]   Posner, F. L., "Spiky sea clutter at high range resolution and very low grazing angles," *IEEE Trans on Aerospace and Electronic Systems* Vol. 38, No. 1, 2002, pp 58-73.

[9]   Finn, M. V., et al., *Uncluttered Tactical Picture*, Metron Inc Report to Office of Naval Research, 17 April 1998. (http://lib.stat.cmu.edu/general/bmtt.pdf).

[10] Skolnik, M., *Introduction to Radar Systems*, 3rd Ed., New York: McGraw-Hill, 2001.

[11] Ristic, B., S. Arulampalm, and N. Gordon, *Beyond the Kalman Filter*, Norwood, MA: Artech House, 2004.

[12] Kopec, G. E., "Formant tracking using hidden Markov models and vector quantization" *IEEE Transactions on Acoustics, Speech, and Signal Processing*, Vol. 34, No. 4, August 1986, pp. 709-729.

[13] Streit, R. L. and R. F. Barrett, "Frequency line tracking using hidden Markov models," *IEEE Transactions on Acoustics, Speech, and Signal Processing*, Vol. 38, No 4, April 1990, pp. 586-598.

[14] Bethel, R. E., and R. G. Rahikka, "Optimum time delay detection and tracking," *IEEE Transactions on Aerospace and Electronic Systems*, Vol. 26, No. 5, September 1990, pp. 700-712.

[15] Lindgren, R., and L. Taylor, "Bayesian field tacking" *Proceedings SPIE* Vol 1954, Signal and Data Processing of Small Targets, 1993, pp. 292-303.

[16] Boers, Y., and J. N. Driessen, "Particle filter based detection for tracking," in *Proceedings of the American Control Conference*, June 25-27, 2001 Arlington Virginia.

[17] Boers, Y., and J. N. Driessen, "A particle filter based detection scheme," in *IEEE Signal Processing Letters*, Vol 10, No. 10, 2003.

Appendix

Gaussian Density Lemma

This appendix presents a lemma concerning densities of multivariate Gaussian distributions that is used to compute conditional densities a number of times in the body of this book. The lemma is also useful in computing association likelihoods involving Gaussian distributions.

We use the notation $Z \sim \mathcal{N}(\mu, \Sigma)$ to indicate that Z has a Gaussian distribution with mean μ and covariance Σ. Let

$$\eta(x, \mu, \Sigma) = \text{the density function for } \mathcal{N}(\mu, \Sigma) \text{ evaluated at } x$$

GAUSSIAN DENSITY LEMMA

Lemma. Suppose

$$
\begin{aligned}
X &\sim \mathcal{N}(\mu, \Sigma) \\
Y &= HX + \varepsilon \\
\varepsilon &\sim \mathcal{N}(0, R)
\end{aligned}
\tag{A.1}
$$

where X is l-dimensional, Y is r-dimensional, H is an r-by-l matrix, and ε is independent of X. Let

$$
\begin{aligned}
f(x, y) &= \textit{joint density for } X \textit{ and } Y \\
f(y \mid x) &= \textit{conditional density for } Y \textit{ given } X = x \\
f_1(x) &= \textit{marginal density for } X \\
f_2(y) &= \textit{marginal density for } Y
\end{aligned}
$$

Then

$$\int f(x,y)dx = \int f(y|x)f_1(x)dx = f_2(y) \tag{A.2}$$

$$f(y|x) = \eta(y, Hx, R) \tag{A.3}$$

$$f_2(y) = \eta\left(y, H\mu, H\Sigma H^T + R\right) \tag{A.4}$$

$$\Pr\{X = x| Y = y\} = \eta(x, \mu_1, \Sigma_1) \tag{A.5}$$

where

$$\mu_1 = \mu + \Sigma H^T \left(H\Sigma H^T + R\right)^{-1}(y - H\mu)$$

$$\Sigma_1 = \Sigma - \Sigma H^T \left(H\Sigma H^T + R\right)^{-1} H\Sigma$$

Proof. Equation (A.2) follows from the definitions of joint, conditional, and marginal densities. Equation (A.3) follows from $Y = HX + \varepsilon$. Since Y is a linear combination of normal random variables, it is itself normal. From (A.1) we can calculate

$$\mathbf{E}[Y] = H\mu \text{ and } \mathbf{Var}[Y] = H\Sigma H^T + R$$

Thus

$$Y \sim \mathcal{N}\left(H\mu, H\Sigma H^T + R\right)$$

and (A.4) follows. To obtain (A.5) we note that

$$\begin{pmatrix} X \\ Y \end{pmatrix} \sim \mathcal{N}\left[\begin{pmatrix} \mu \\ H\mu \end{pmatrix}, \begin{pmatrix} \Sigma & \Sigma H^T \\ H\Sigma & H\Sigma H^T + R \end{pmatrix}\right]$$

Then (A.5) follows from a standard result on conditioning multivariate normal distributions. See Property 3.3 in Section 3.4 of [1].

References

[1] Morrison, D. F., *Multivariate Statistical Methods*, 3rd Ed., New York: McGraw-Hill, 1990.

About the Authors

Lawrence D. Stone received his B.S. in mathematics from Antioch College and his M.S. and Ph.D. in mathematics from Purdue University. He is the chief scientist for Metron Inc., where he has worked since 1986. Prior to Metron, he worked for Daniel H. Wagner, Associates. Dr. Stone has been involved in numerous underwater search operations, including the search for the U.S. nuclear submarine *Scorpion* lost off the Azores in 1968, the SS *Central America* lost off the coast of South Carolina in 1857 while carrying gold and passengers from California to New York, and the Air France Flight 447 lost over the Atlantic in June 2009. He is one of the designers of the U. S. Coast Guard's Search and Rescue Optimal Planning System. His book *Theory of Optimal Search* was awarded the Lanchester Prize by the Operations Research Society of America. Dr. Stone is a member of the National Academy of Engineering, a fellow of the Institute for Operations Research and Management Science, and a recipient of the J. Steinhardt Prize for outstanding contributions to military operations research. His current technical activities include the development of multiple target detection and tracking systems for U.S. Navy.

Roy L. Streit received his B.A. (Honors) in physics and mathematics from East Texas State University, his M.A. in mathematics from the University of Missouri, and his Ph.D. in mathematics from the University of Rhode Island. He was a visiting scientist at Yale, 1982–1984, and a visiting scholar at Stanford University, 1981–1982. He is a senior scientist at Metron. Prior to joining Metron in 2005, he was a senior scientist in the senior executive service at the Naval Undersea Warfare Center, working on multisensor data fusion algorithms for submarine sonar and combat control automation. He received the Solberg Award from the American Society of Naval Engineers, and the Superior Civilian Achievement Award from the Department of the Navy. He was president of the International Society for Information Fusion in 2012. Dr. Streit is a fellow of the IEEE. His book *Poisson Point Processes: Imaging, Tracking, and Sensing* was published by Springer in 2010 and in Chinese in 2013. He holds nine U.S. patents. He is a professor (adjunct) of Electrical and Computer Engineering at the University of

Massachusetts at Dartmouth. His research activities include multitarget tracking, medical imaging, and statistical methods for pharmacovigilance and business analytics.

Thomas L. Corwin received his B.S. in mathematics from Villanova University and his M.S. and Ph.D. in statistics from Princeton University. He is currently serving as President, Chief Operating Officer, and Board Chairman of Metron, Inc., which he founded in 1982. Prior to that he worked for Daniel H. Wagner, Associates. Dr. Corwin has been involved in search operations that include operational military searches, searches for downed aircraft, a U.S. Coast Guard effort to recover 500 barrels of cyanide lost in the Gulf of Mexico, and the removal of unexploded ordnance from the Suez Canal following the Yom Kippur war. In work for the Commander Submarine Forces, U.S. Pacific Fleet, he designed computer systems to search for and track submarines. For this work, he received the Secretary of the Navy's Distinguished Public Service Award. Dr. Corwin's recent research activities include the adaptation of the methods of statistical physics to large scale, many body inference problems.

Kristine L. Bell received her B.S. in electrical engineering from Rice University, her M.S. in electrical engineering from George Mason University (GMU), and her Ph.D. in information technology from GMU. From 1996 to 2009, she was an associate/assistant professor in the statistics department and C4I Center at GMU. During this time she was also a visiting researcher at the Army Research Laboratory and the Naval Research Laboratory. She joined Metron in 2009, where she is a senior scientist. She also holds an affiliate faculty position at GMU. She is the coauthor (with Harry Van Trees) of the second edition of the book *Detection, Estimation, and Modulation Theory,* published in 2013 and coeditor (with Harry Van Trees) of the book *Bayesian Bounds for Parameter Estimation and Nonlinear Filtering/Tracking,* published in 2007. In 2009, she received the Outstanding Alumnus Award from GMU's Volgenau School of Engineering. Her research activities are in the areas of statistical signal processing and performance bounds for source localization and tracking with applications in radar, sonar, aeroacoustics, and satellite communications.

Index

Signal Detection and Estimation, Second Edition, Mourad Barkat

Signal Processing in Noise Waveform Radar, Krzysztof Kulpa

Space-Time Adaptive Processing for Radar, J. R. Guerci

Special Design Topics in Digital Wideband Receivers, James Tsui

Theory and Practice of Radar Target Identification,
    August W. Rihaczek and Stephen J. Hershkowitz

Time-Frequency Signal Analysis with Applications, Ljubiša Stanković,
    Miloš Daković, and Thayananthan Thayaparan

Time-Frequency Transforms for Radar Imaging and Signal Analysis,
    Victor C. Chen and Hao Ling

For further information on these and other Artech House titles, including previously considered out-of-print books now available through our In-Print-Forever® (IPF®) program, contact:

Artech House	Artech House
685 Canton Street	16 Sussex Street
Norwood, MA 02062	London SW1V HRW UK
Phone: 781-769-9750	Phone: +44 (0)20 7596-8750
Fax: 781-769-6334	Fax: +44 (0)20 7630-0166
e-mail: artech@artechhouse.com	e-mail: artech-uk@artechhouse.com

Find us on the World Wide Web at: www.artechhouse.com

DONALD TRUMP
IS NOT MY SAVIOR

DONALD TRUMP

IS NOT MY SAVIOR

AN EVANGELICAL LEADER
SPEAKS HIS MIND ABOUT THE
MAN HE SUPPORTS AS PRESIDENT

MICHAEL L. BROWN, PH.D.

DESTINY IMAGE® PUBLISHERS, INC.
P.O. Box 310, Shippensburg, PA 17257-0310
"Promoting Inspired Lives."

This book and all other Destiny Image and Destiny Image Fiction books are available at Christian bookstores and distributors worldwide.

For more information on foreign distributors, call 717-532-3040.

Reach us on the Internet: www.destinyimage.com.

ISBN 13 TP: 978-0-7684-4993-8

ISBN 13 eBook: 978-0-7684-4994-5

ISBN 13 HC: 978-0-7684-4996-9

ISBN 13 LP: 978-0-7684-4995-2

For Worldwide Distribution, Printed in the U.S.A.

1 2 3 4 5 6 7 8 / 22 21 20 19 18

CONTENTS

CPSIA information can be obtained
at www.ICGtesting.com
Printed in the USA
LVHW081327051218
599360LV00017B/119/P